POLLUTE
AND BE DAMNED

POLLUTE
AND BE DAMNED

Arthur Bourne

with illustrations

LONDON
J M DENT & SONS

First published 1972

Made in Great Britain
at the
Aldine Press · Letchworth · Herts
for
J. M. DENT & SONS LTD
Aldine House · Bedford Street · London

ISBN: 0 460 03980 6

CONTENTS

LIST OF ILLUSTRATIONS

Plates 1, 2, 4, Arthur Bourne; Plates 3, 9, 10, USIS; Plate 6, *Italia Nostra*; Plate 7, Thor Heyerdahl; Plate 8, Eric Perkins: Plates 11–14, NASA.

PREFACE

This book is an attempt to put the problem of pollution into perspective, a problem which, along with most of the other problems that beset us in this latter half of the twentieth century, is the result of our mis-management of ourselves and the Earth's resources. It is a problem which so often seems to be beyond our capability to do anything about and even to understand, but as Albert Einstein once said, 'The most incomprehensible thing about the world is that it is comprehensible.' If we can comprehend the problem of pollution, then we are at least some way towards its solution, and the pages that follow are an effort in this direction.

The author of this book owes much to others. Wherever possible I have given the names of the original workers and sources of information in the text, in addition there are many whom I do not know and to these I am grateful. My thanks must also go to those who have borne with me, helped and criticized my work. Sometimes I have taken their advice, sometimes I have decided against it, the facts are everybody's, the way they are presented and the opinions are mine.

I am particularly grateful to my parents who encouraged my curiosity about the world I had been born into; to the person who has the most difficult job of all, that of being my wife, and to those who have heard it all before, Julian, Linda and Anna; to Dick Fifefield (Managing Editor of the *New Scientist*) who was instrumental in bringing together an author in search of a publisher and a publisher in need of an author; to J. M. Dent & Sons, my publishers, for their patience; to Dr F. Steele without whom this book probably would not have seen the light of day; to Dr C. Maghrabi and Dr S. Frank for research and encouragement; to Dr E. Perkins for reading Chapters 8 and 9; to Dr G. Fielder for reading Chapter 11, and Brian Aldis for reading Part I in its entirety and wisely counselling me against overdoing the Bible thumping; to Dr T. Gaskell for helpful suggestions, and to Lyn Jenkins for working so hard

at such short notice to prepare the manuscript; to Jeremy
Weston and his staff of the Library of the Royal Institution;
to Bill Dunn of the United States Embassy Press Office for
his assistance in obtaining photographs and background
material; to Dr Robert de Ropp for allowing me to use the
extract on page 134; and finally to Jeni Couzyn for permission
to publish her new poem.

January 1972   A. G. B.

There Are Some Creatures Living In My Body

There are some creatures living in my body. I bid them
Welcome. Let them feed off me, as I off wild creatures
                                    that run free.
Let my veins and bones be to them rivers and baobabs, let
cells be huge rich valleys, let gigantic landscapes
roll and change as I flex my nerves.
O I wish them an excellent universe, such a one
as I inhabit, mountains and wind and
a lot of stars. Nor let them
pollute and destroy what they find—let my rivers of
                                    blood
flow clean, my flesh be fertile and multiply, nor cloud
                            with stale chemicals
the clear windows of my eyes.

Jeni Couzyn

PART I

GENESIS

CHAPTER 1

A problem of parts, of a problem

A man stumbled off a tube train, tottered a few steps along the station platform and then collapsed in a paroxysm of coughing. His sputum was tinted pink. Four hundred miles away in another city a doctor working in a hospital renal unit died of infective hepatitis. On a Shropshire farm a rodent officer, unable to control poison-resistant rats, packed his bags and drove away. On a coast a few miles to the west lay thousands of dead and dying seabirds. In another country forty million fish floated belly-up, down a once beautiful river. Half the world away in the blue ocean a starfish changed its habits and became a plague.

These incidents are not imaginary, they have all happened in recent months—and what is more, they are all connected. Most of them made headlines in the daily newspapers. They are all aspects of contemporary life. They are parts of a problem—the problem of pollution, which is itself only part of an even greater problem. Pollution, along with the over-exploitation of resources, shortages of food and water, vanishing wildlife and vegetation and social decay, is the result of the over-burdening of the Earth with human beings.

These problems had their birth during that period referred to as the Neolithic revolution, when Man was changing from hunter to agriculturalist. While he remained a predator his effect on the environment was no more and no less than that of any of the other animals then living in predator-prey relationships. Then he made a profound discovery—he found out how to domesticate certain plants and animals, and for these he had

1

to clear the land of its natural vegetation and drive off or kill the other animals. Man began to change the environment. He had started on his conquest of nature. At first he was able to clear only little patches of land for his own use and when these were worked out he moved on. The old patches were soon recolonized by the wild plants and eventually trees grew again, but Man had made his mark. While his activities were so confined no great damage was done, for the energy cycles were only temporarily redirected into his less efficient systems, short term economical systems which are at the root of most of our troubles today, not only in agriculture but in all areas of industry too.

This early agriculture provided a surplus of food which could be stored for use during the winter or minimal growing periods and to see the community through times of crop failure. It also provided sufficient for trade with other communities and it enabled Man to diversify and to specialize in the various things we now know as art, science, law, politics, etc. From that time onwards the population of the human species began to increase more rapidly. In common with all other animals, the numbers of the human race were dependent on the availability of food. In those early stages of development, and indeed until quite recently, famine, disease and war kept the population in line with the food supply. But the advent of specialization enabled Man to devise ways of preserving food so that, in the more successful communities, famine became less effective as a means of population control.

Gradually a degree of mastery of his environment gave Man great advantages and his numbers steadily increased. In ten thousand years he effectively changed the environment to suit his needs. The refuse of his early years forms the treasure of the archaeologist today. He discovered that he could increase the productivity of his crops by irrigation which in some climates enabled him to reap more than one harvest a year. Irrigation, however, also increased the area of water for disease-carrying organisms, particularly the malaria-carrying *Anopheles* mosquito. From that time, up to the twentieth century, malaria became a major disease of mankind and one of the most effective checks on his numbers. Nature was desperately trying to balance things out—as Man increased his food supply, she countered with disease.

Slow as they were, the changes were being made; the energy was beginning to be redirected through a food chain that led only one way, with little being returned to maintain the equilibrium in the habitat. The so-called Industrial Revolution of the eighteenth and nineteenth centuries speeded up the process, with the addition of massive pollution of air and water. If the environmental problems had remained at the level they had reached prior to, say, the tenth century, nature might have been able to cope with them. But the growth in mining, quarrying, burning of coal and use of rivers as sewers, coupled with the rather sudden increase in the human population, brought about severe and fundamental changes on the land.

Man had already denuded vast areas through his uneconomical agricultural methods, and particularly through the grazing of his domesticated animals. There is little left of the wild places of ancient Britain, except perhaps the very tops of the high mountains. Most of what we think of today as the 'wild' moors and fells are man-made—produced by tree felling and over-grazing. Similar regions can be found all over northern Europe; even Iceland owes much of its barren interior to over-grazing; while in the Mediterranean countries these activities have left stony barren lands like those of the Aegean. China's enormous expanses of poor soil are the result of the destruction of her forests for agriculture and industry over three thousand years ago. Today large areas of China have no water reserves and her great rivers carry more silt to the sea than all the rivers of the world put together. The Chinese have many times tried to reintroduce forests, but each time the demands of a growing population have defeated them. In the arid countries of North Africa and the Middle East the same activities have produced the familiar sand and rocky deserts. Almost the whole of the

Sahara and the deserts of Arabia were once fertile. Parts of North Africa were covered with immense tracts of forests, and as recently as the time of Carthage small forest elephants roamed the area; it is this species of elephant that some zoologists believe Hannibal used in his march against the Romans during the Second Punic War. Even today the Sahara is increasing by about ten thousand acres a year, despite attempts to arrest the spread and to bring some areas under cultivation again.

Man had become a prime factor in the environment with almost, but not quite, equal status to the physical forces which promote change or maintain stability in the habitat. The growth of the human population began to increase at an exponential rate, so much so that during the last twenty-five years the line on the graph has become almost vertical. There are 3,500 million human beings alive today—twelve times what there were on the first Christmas Day. Three-quarters of this rise has taken place since the middle of the eighteenth century. In the early communities the numbers of people were important; the larger the population the greater the security—military strength was dependent on the number of fit males available for war. The Industrial Revolution created a new need for manpower; men and women and even children were required as factory workers, and the larger the family the more the family income. Forced or attracted to the booming industrial towns by the promise of work throughout the year and relatively high wages, the peasantry left the land. People were encouraged to have large families. The industrial barons needed people to make and, equally important, to buy the products of the factories—a need being perpetuated today, another cause of our twentieth-century predicament.

The tall factory chimneys belched their foul fumes into the atmosphere and the wastes of the mills and foundries were poured 'willy-nilly' into the rivers; add to this the smoke from the clusters of dingy workers' houses and you have all the elements that made so many of the industrial towns the unhealthy places they were. Man had created a new habitat: the industrial community—the beginning of a new Hell on Earth. England, where this all began, became prosperous, her manufactures were sold all over the world and she could afford to buy not only any necessary raw materials—she didn't need many, she had most of her own—but food and even luxuries; she could create an empire upon which 'the Sun never set'. The open-ended energy system so alien to nature was launched. Waste was money, 'where there's muck there's brass' became a familiar slogan—to some, it became a motto. What was happening to the landscape and rivers did not matter to these new, self-made men; if land was not productive then it must be made to be productive. Industry reproduced itself at a rate, for those days, beyond belief—like some malignant cancer it lived on the land, the water and the people—it absorbed and eventually ruined them. Man had lost sight of nature, and of his own in particular. His success had been so quick and seemed so absolute that he believed in his superiority—a thing apart. Nature at last was beaten.

The revolution spread from England to Europe and the United States of America, where it was to be given a new and greater power: mass production, throwaway produce, 'built-in obsolescence', the principles of modern industry and commerce. Waste, dereliction and pollution grew as the population increased. The eradication of many killing diseases and their insect carriers, especially after the Second World War, removed the remaining restraints on population growth, other than war. The non-industrial countries began their spiral to self-destruction—mounting populations that poor agricultural methods could not feed. People were, and still are, being saved from death by disease to face a future of subsistence diet, starvation and early death. Cities and towns spread out and obliterated the old villages and agricultural land. The demand for food in the more prosperous industrial states brought forth intensive farming and the need for artificial fertilizers and an armoury of chemical pesticides. Man seemed to be master, in reality he was only a short step ahead of nature. Man had invented and lived in a 'one-way energy system'.

Too many people, too little food, decreasing resources, mounting wastes, derelict land and poisoned water and air are the legacies of the system. It has been a long journey over the past ten thousand years or so and, just when it seems that Man has accomplished all that he could wish for, nature has begun to take it all away. The man dying of lung cancer, the doctor struck down by a virus, the immune rat, the dead birds, and fish, and the menacing starfish are all symptoms of the illness the world is suffering from today. Man has gone on congratulating himself for too long; he should have known better. Being an animal, he could not hope to escape indefinitely the biological backlash. Pollution is a problem of parts, but it is only part of the greater problem, that of too many people.

CHAPTER 2
A matter of numbers

The enormous number of people living today is evidence of Man's success in his conquest of nature. It is the triumph of modern medicine which has provided him with a means of death control. Populations increase when the birth rate is greater than the death rate, whereas in a balanced population these rates cancel each other out. A normal population increases by geometrical progression; bacteria provide a very simple illustration of this. Bacteria reproduce by division, each individual divides into two daughter cells, each of which in turn divides into two and so on. Thus one bacterium can in the course of hours produce millions of descendants. A man may take years instead of hours, but he is more prolific, in the sense that he does not simply divide but reproduces sexually and can go on reproducing and still survives as an individual. He reproduces children for many years. A man of ninety years could have two hundred descendants living during his own lifetime. At the present rate of increase—two a second—the human population of the Earth will by the end of the present century be coming up to 7,000 million. That is an increase of 6,000 million in 170 years; it took about a million years to reach the first 1,000 million. The climbing rate of population growth does not mean that women are having more children *per se*, rather it is a reflection of an increase in the survival rate of infants and young children and of the fact that people are living and breeding longer. Another striking phenomenon relating to mortality is that the pattern of death has changed.

7

If we ignore technological injury and death there are two kinds of mortality: environmental death, which today more commonly applies in the under-developed countries, and degenerative death. Environmental death is caused by infectious diseases, malnutrition, trace element and vitamin deficiencies, and natural catastrophes such as earthquakes and volcanic eruptions. Degenerative death includes mortality due to conditions such as heart disease, cancer and related biochemical imbalances. Death from these is becoming more common in industrialized and urban communities. It may be that under a more natural regime people who die from these causes would have died earlier through one or other of the environmental causes of death—primitive Man rarely lived beyond his thirtieth birthday—but because of modern medicine, sanitary conditions and the availability of food they are living longer.

In former times famine and disease afforded the real biological checks to the human reproductive rate—the Black Death killed off a third of Europe's population, and starvation nearly decimated Iceland's in the eighteenth and early nineteenth centuries. War has become, in spite of the millions slaughtered in the two world wars, less effective as a population controller, unless, as we shall discuss later, nations at some time in the future resort to all-out biological or nuclear warfare, which would be tantamount to biocide anyway. In other animals, when the population reaches a density that stretches the food supply to its limit, and the risk of breakdown is imminent, several mechanisms come into play to reduce the reproductive rate: for instance, embryos and foetuses already developing in the females may be reabsorbed.

In controlling many of the causes of environmental death Man is unwittingly bringing upon himself the very real possibility of his own extinction in the long term. To meet the demands of the increasing population, food productivity has to be stepped up and this can be achieved only by making severe inroads into the natural ecosystem. It takes a long time for any major natural alteration in the environment to show itself. Tremendous inertia tends to preserve the *status quo*, but already there is evidence that Man's activities, particularly since the First World War, have begun to move the immovable. Changes in the atmosphere, water regime, and natural food cycles are not hard to discern. Even so, it is unlikely that Man will completely alter these regimes; on the contrary, it is more likely that breakdowns in the systems will occur first and result in stoppages in the energy chains supplying the basic necessities of life.

Whatever happens, there can be little doubt that the physical forces which operate on this planet will shift the scales to restore a cycling energy system, which is a natural one and obeys the laws of the physical and chemical universe. Man could avert disaster by applying ecological principles *now*, but this would require a drastic rethink on his part, and above all on the part of those who take it upon themselves to lead and govern communities. The short term economics, whether they be in agriculture or industry, will have to be abandoned. This will probably mean hardship and a lowering of affluence in wealthy countries. It will mean control of human reproduction, and though an unpopular step, compulsorily if necessary.

Today it is idle to speculate too far into the future. We can expect, in the short term at least, a continuing decline in mortality rates and, in the 'have' countries, an increase in the success of medical science. The horror of living hundreds deep and in a world totally covered by concrete and steel suggested by some writers will never happen. Long before a situation even vaguely resembling that state is reached, social decay will have brought civilization to a state of collapse, or society will have hardened and the opposite, oppressive dictatorship—reminiscent of Orwell's *1984*—will have brought a halt to the kind of civilization most people would prefer to live in.

Mankind does not have to wait for the changes caused by over-population, whatever that may mean, to overtake him. He has enough problems already. The human species is already

too numerous for its own comfort and the continuance of the
present trend will produce a multiplication of the environmental
damage even before the 1970s are out. Countries with high
population densities such as the United Kingdom and Japan,
will suffer most. Japan has an enormous problem already, but
the United Kingdom is a good model to examine. It is an
example of a country that has an excessive population—one
which the environment cannot support—a state reached a long
time back. Some estimates have put the carrying capacity of
the United Kingdom at 25 million; we have over twice that
number today and by the end of the century it will have increased by another 40 per cent. Whatever the number, most
agree that the United Kingdom has passed its optimum population and the situation can only decline. In practical terms it
means that Britain has to trade its technology for its food and
raw materials; we exhausted most of the indigenous ores years
ago. It means increasing unemployment and poverty. If a man
is not a member of the better-off classes or organized society
he has little chance of survival thirty years from now. Excessive
population means unwanted people, especially the old, dependent on society for sustenance; it means 60,000 deaths for the
price of keeping warm; it means 400,000 without homes;
it means an increasing gap between the haves and the have-
nots; it means more waste and more pollution, and it
means less money to combat this problem. It means
a country on a knife-edge—a walk with disaster. The United
Kingdom, regardless of political colour, will have experienced
acute hardships and shortages long before this decade is out—
the longer-term view is dependent on the next ten years. And
although during the last few years we have had signs enough,
there are still those who doubt the dangers.

CHAPTER 3
Pollute and be damned!

The attitude of the average citizen towards pollution is one of incredulity, a disbelief in the gravity of the situation. This has been caused partly by a succession of warnings of impending doom. Cries of 'wolf' have become commonplace, and the question which is being asked now is 'Why should the prophets of doom of 1972 vintage be listened to any more than those of the past?' And it is a reasonable question to ask.

I would have asked the same question myself nearly twenty years ago when we heard that the Americans were to test a new kind of atomic bomb. I can still vividly remember sitting at a laboratory bench waiting for the news and results of that first atmospheric thermo-nuclear detonation. Rumour had been rife, and even in the citadels of science there was talk of a world catastrophe; some people thought the explosion of a hydrogen bomb would trigger off a chain reaction in the atmosphere that would in a few hours destroy us all. Well, it didn't happen, and several detonations later we are still living; but— many have died, many more of us are dying and will die through the enormous injections of radioactive material into the atmosphere from those test explosions. Those who died have been, and those who are dying will have been, murdered by the nuclear powers without a declaration of war, and this is the point: it is not always the obvious that catches us out in the end. There is a fable in Aesop, in which a rich man dreams that his only son is to be killed by a lion. To protect his boy the father has him confined to the house, but when one day the

11

youth sees a picture of a lion in his father's picture gallery, he strikes at it out of sheer frustration and catches his hand on a nail in the wall. The wound becomes infected and eventually the boy dies. Thus the prophetic dream is realized but not in the way the old man expected.

So it is with the slag heaps, spoiled rivers, rubbish dumps and polluted air. These are so much part of our habitat that many of us have been brought up with them so as to regard, or rather disregard, them as the normal scheme of things. The living of our everyday lives leaves little time to think of anything but the really unusual, so it is hardly surprising that the last place we look for danger is right on our own doorsteps. We have not been aware of any risk to the environment, and even if we had been, I doubt very much whether many of us would have known what to do about it. But the risks have been there, slowly, subtly, but relentlessly building up to danger point until they have forced themselves on our consciences and can no longer be ignored. Large numbers of dead birds in Lincolnshire caused a flutter amongst ornithologists, the massive sea-bird kill in the Irish Sea jolted many more of us; the realization that the sea is polluted gradually dawned. People began to sit up and take notice—thalidomide babies, nerve-gas injuries, dead sheep, seals, fish; could they perhaps be part of a pattern, not just a series of isolated incidents?

But, with all the evidence to show that all is not right, there are still far too many people who are perpetuating and intensifying the danger. The early industrial masters, obsessed with the overwhelming drive to go on producing with no heed for the consequences may, perhaps, be partly excused for their 'produce and be damned' attitude because it was born of ignorance of the wider issues, but there is no excuse for having that attitude today. There is no hiding behind ignorance, there is too much evidence, and yet there are still many people (not all of them industrialists either) whose attitude is the same as that held by the founders of the Industrial Revolution. It is a way of thinking, pervading society, based on the overriding human propensity for selfishness and greed. No one member of the human race is entirely devoid of selfishness, and with the increasing population pressure the 'every man for himself' attitude is more prevalent than ever. After selfishness come apathy and a total disregard for the future.

Danger has to be tangible before people start caring about what, to them, is an abstraction, and to most people conservation, and even the environment, are abstractions. They hear of them mostly from 'clever' people, somewhere else, usually with full bellies and living in comfortable circumstances. They do not believe that they can be under any threat from pollution, the lack of resources or anything else for that matter—it always happens to the other guy! Meanwhile they go on in the same old way, polluting and degrading the habitat as their fathers and grandfathers did before them. Many people, particularly industrialists and politicians, take the line that the ends justify the means. Even in 1970, the so-called European Conservation Year, many of those who paid lip service to the conservation and anti-pollution lobby were still defending this attitude, as was abundantly clear during the many conferences that were held.

Pollution is a vast problem and is not confined to any one country or group of countries. It is truly global and recognizes no boundaries. The size of the problem certainly seems to be beyond the control of any one person, group of people or even nations. It must be seen as a world problem, but this should not be used as an excuse for doing nothing. If pollution is a consequence of population numbers, then each one of us is a polluter, all 3,500 million of us. Admittedly the vast majority contribute only a very small fraction to the problem, for the main polluters are the citizens of the industrialized countries where the standards of material prosperity, at least, are very high. The peoples of North America, Europe, the Soviet Union and Japan are thousands of times more responsible for the environmental crisis than are the peoples of the under-developed countries.

Among those who ought to know better we find the same kind

of attitude, stubbornness that can be matched only by their limited imagination. For example, in the United Kingdom there are still leaders and members of local councils who refuse to accept or believe in smokeless zones despite the fact that atmospheric pollution has been shown to be responsible for the sickness and deaths of thousands of people. Furthermore many owners of industrial plant have taken it as a right to use streams and rivers for the disposal of their wastes, and local authorities dump their sewage, treated or untreated, in them as a matter of course. When this gets difficult there is always the sea, a seemingly bottomless cesspit. Thousands of tons of industrial wastes are daily taken out to sea and dumped. The accepted way of getting rid of radioactive wastes is by 'burial' at sea. Nobody much cares what happens to these, providing they don't show up again in our lifetime. What happens a hundred, a thousand or forty thousand years hence is not our concern: we pass the problem on to the next and subsequent generations.

Our thinking is too parochial, our approach too fragmented, so much so that it distorts the problem and makes it more difficult to solve. The image most of us have of what is meant by the word 'environment' is of the countryside with its trees and meadows, or of our immediate surroundings. Whereas in reality the environment is the sum total of all those factors that make up the world we live in: biological, physical, chemical, social and psychological. It is like a jigsaw puzzle: some of the pieces are small and some are large, but all are vital to the picture. Some pieces are the key to the whole—lose one belonging to a jigsaw puzzle and the picture does not make sense; lose one belonging to the environment and the whole picture disintegrates.

Some parts of the environment are more vulnerable to change than others—especially quick changes, and Man's are on the whole quick. It is convenient to break down the environment into units called 'habitats', or further into the physical and biological factors of which it is composed, like a chemist breaking down a complex chemical compound into its constituent atoms. But just as the chemist cannot understand the compound by studying its atoms separately, so the environment cannot be understood wholly by study of its separate aspects. In the past many people considered nature as distinct from Man, and as a result many erroneous ideas were extant. Naturalists studied the animals and plants of the countryside and ignored Man. More recently conservationists have repeated these errors and tended to leave Man out of their calculations, except as a destroyer of habitats and a butcher of wildlife. Yet our houses, schools, factories, roads and railways are as much part of our environment as are the mountains and sea. Few of us in the industrial countries can remember a landscape without these features and most of us would perish without them. The troubles that confront us now are not due to our ability to build houses and factories, etc., or to our cleverness in extracting ores from the earth or in damming the streams and rivers. They arise because we do not give enough thought to the wider consequences of what we do to the environment; we compartmentalize our activities so effectively that there is little exchange between the compartments.

The 'separateness' approach applies in everything Man does. It is often said that there is so much to know that no one person can take it all in, but it is doubtful whether we should act any more wisely if we could. Our individual goals would favour one thing or another at the expense of something else. The result of this has been that, with few exceptions, there has been little care for the environment as a whole and precious little attention given to the psychological, sociological and physical effects that our activities have, and are having, on the members of the community.

Just as the past gave rise to the present, what we do today will affect tomorrow, and the 'I'm all right, Jack' attitude is perhaps the greatest threat that faces Man and his environment. It already threatens the economy of Britain and many other countries. But it goes beyond the present, it threatens the very survival of mankind. Consequently we must expect the growing pile of wastes, degradation of the environment and psycho-social disorders of society to increase, at least for the time being. The whole fabric of Western society is heaving with unrest, and violence breaks out at the least provocation. Before very long it will explode and a lot of people are going to get hurt. From somewhere inside the ferment voices have been heard, warning that all is not right, yet those enjoying the boom of the commercial revolution are sceptical of these warnings. A few thousand birds die in the Irish Sea, 40 million fish perish in the Rhine today, a hundred thousand people starve to death tomorrow—does it matter?

Collectively and individually we are responsible for this planet of ours and, if we wish, we can still put things right. But perhaps we do not know the extent of the damage. Let us look at the problem and see what can be done about it. If we continue with an attitude of pollute and be damned, the tragedy is, we *shall* be damned!

PART II

MAN IN AN ENVIRONMENT

CHAPTER 4
Manhive

In a sense, many of us are already damned, trapped in the stinking, noisy workings of our machinery. All the attitudes, all the errors of judgment are to be found in concentrated form in that most human of all human phenomena—the city. Half the world is dependent on cities. The city-dweller, immersed in the machinery of commerce and industry and enclosed in the canyons of artificial caves, lighted, heated, watered and fed, is little better off than the rats in the scientist's laboratory. He is unable to exist without the city: urban Man, more than any other, is prone to forget his dependence on nature and it is in urban communities above all that we find a *laissez-faire* attitude to the environmental problems.

Many of contemporary Man's ailments, his hypertensions and neuroses, are the result of his divorce from his natural environment and his frustrations in the stifling conglomerations that he has built around himself. We do not need the researcher's rodents to tell us that we are under stress—we feel it every day, we feel it when we wait in traffic jams or sit in a stranded train, we feel it when the noise of our surroundings overwhelms us. We live with it, hating it, until in the end it is capable of destroying us.

The city absorbs most of our energy, it sucks the life from a nation's people. It spreads, sending out tendrils of concrete and metal; it joins, absorbs or competes with others of its kind for the necessities of its existence. It is man-made but has a malignant will of its own. It owes no allegiance, has no natural

17

order and its boundaries are limited only by the resources that feed it.

The large modern city has in many cases been dehumanized. From being the cultural and spiritual heart of a people, it has turned into the commercial and industrial centre of an urban sprawl. No longer does it afford unity and security to its citizens, it threatens them with physical and mental violence. It is in the cities of the twentieth century that the seeds of the collapse of industrial civilization have been sown. It is an environment largely ignored by the conservationists and the ecologists, and yet it is in the city that the contemporary problems that affect Man, and ultimately the world environment, have their roots. There are about a thousand cities in the world with populations of over 100,000 people, and by the end of the century there will probably be another thousand: another thousand centres of congestion, of misery and shortages. The environmental problems are intensified in the metropolis, the individual finds himself alone and afraid, unable to come to terms with his surroundings. He feels the lack of the bonds that held the family, the tribe, the group together in earlier times. He feels listless and without roots; he no longer 'belongs'. Landmarks that once defined his territory are obliterated by row upon row of man-made cliffs. The supermarket or multi-storied car-park has replaced the cathedral as centrepiece. The holy place was built for eternity; buildings today are built for today—there is no tomorrow! Familiar skylines become blocked by a palisade of monoliths. London is being destroyed by the developer: to him the open spaces are irritatingly non-productive. The elegance of St Paul's is lost amid gargantuan mediocrity, and the 'Central Parkization' of Hyde Park has begun. Even this great city is becoming dehumanized. It is developing into a metropolitan Frankenstein.

The effects of all this on the human mind are only just being guessed at, but already in terms of unhappiness they are more than a match for all the other pressures. Even so, people are still pouring into the cities from rural areas, bringing with them their needs and adding to the social and physical problems. With every week that passes the difficulties are getting more complex. In these circumstances it is not surprising that the average citizen cares little for the environment. It seems to be completely out of his hands. His senses are dulled; he accepts the scheme of things as inevitable, for he is unaware of any alternative; but should he become aware, he is uncomfortable, for he fears change. It is in this state of apathy that we have allowed our habitat to become degraded. We might congratulate ourselves about our progress in clearing the slums and dereliction which are the legacy from the rush to participate in the early days of the Industrial Revolution, but what do we replace them by? Dwellings, factories and public buildings which are unimaginative in design and sometimes built so carelessly that they begin to decay almost as soon as they are put up. Yet the majority of people accept all this; but is it surprising, when they are wretchedly untidy in themselves? Most British towns, particularly in the Midlands and the North, are littered with Man's droppings—newspapers, sweet-papers, cigarette packets, cigarette ends. The shabbiness and ugliness of our cities are not merely a reflection of the meanness of councils, industrialists and developers, they are a reflection of the citizens themselves.

The city is like a cancer spreading out relentlessly and obliterating everything that stands in its way. Unchecked, the 'megalopolis' of the future will be little more than a cage from which those born in it will never escape. This is not a forecast for thousands of years in the future, it is happening now. The seeds of the new conurbations are being sown all over the countryside of the industrial nations. In Britain the growing points of the gargantua are already on the move. In the last twenty years 750,000 acres of farmland in England and Wales have been used for urban development. London is spreading towards Bristol and Bristol reaches out towards London. Reading and Swindon will be swamped. Southampton grows towards Winchester and will engulf Basingstoke to join ultimately with the London–Bristol complex at Reading, while the proposed city centred on Newbury will be a mere suburb of the 'Lontol' megalopolis. Plymouth threatens to march to Bristol, Manchester and Liverpool to join and cover the whole of South Lancashire and spread to Bradford, Leeds, Sheffield, Chesterfield and Derby. Southwards from Liverpool the giant will join with the Black Country. Birmingham and Coventry will join with Cardiff and Bristol and eastwards reach to Leicester and Nottingham. In the north-east the Durham-Newcastle-Sunderland complex will join with 'Glasburgh' across the 'waist' of Scotland. The whole of Britain will be a formless 'googleopolis'.

It has been forecast that within a hundred years a huge complex of urbanization will stretch from Manchester to Milan and from Madrid to Moscow. By the year 2000 the United States will have at least three megalopolises, 'Boswash' (Boston to Washington D.C.), 'Chipitts' (Chicago to Pittsburgh) and 'San-San'. Boswash will contain a quarter of the population of the whole of the United States, Chipitts will have an eighth, but might also spread to include all the industrial cities of the Toronto region of Canada which would give it a population of over 40 million people. 'San-San', stretching from San Diego to Santa Barbara in Southern California, may join ultimately with San Francisco. The problems inherent in this sort of accumulation will be beyond Man and his technology to solve. The effects on the environment will be colossal, the destruction of the resources enormous and the dehumanization of the Manhive complete. Most of this could come about in less than thirty years from now. Only three things can prevent it happening: man-made or natural catastrophes or the lack of resources.

Cities draw towards themselves all the food and other resources from the land and return very little, except their wastes. To repair and extend themselves they require building materials, stone, clay, sand and gravel. Their demand for water can be met only by taking it from the countryside, sometimes hundreds of miles from the city's boundary, often causing social and political unrest. After all, why should Welsh valleys be flooded to supply Liverpool and Birmingham with water? Why should Derbyshire serve Yorkshire? Why should beautiful Lakeland be spoilt by Manchester? These are valid questions but ones which are answered only in terms of economics and always in favour of the city. In itself the huge mass of an urban area is useless in environmental terms. Rain falling on a city, if it is not evaporated by the heat from the buildings and roads, runs off down the drainage system, often causing surges of water through the sewage systems which make it difficult for the sewage works to cope. Consequently, contaminated water enters the rivers and may be taken into another town's water supply farther downstream. Very little water is absorbed or saved to supply the populace, and the greater the area that is covered by buildings and roads the less water is going to be available.

The growing cities will consume fuels at an ever-accelerating rate. According to world surveys there is about 500 years' supply of coals and lignites that could be mined economically using present-day techniques, and there might be just enough oil to last a little beyond the year 2000. But that is based on the present rates of consumption, and these are not likely to stabilize during the next thirty years. We can see how the rates have accelerated: coal has been mined for about eight hundred years, but about half of the coal produced during that time has been mined since 1930; half of the world's production of oil has been achieved since 1960. It is unlikely, therefore, that fossil fuels will continue to supply much of the energy required by the end of the century. The introduction of nuclear power brings its own problems, the most serious of which, at the moment, is the disposal of the radioactive wastes. If energy can be generated by nuclear fusion instead of fission, there is a chance of a cleaner and a cheaper source of energy, but this seems to be a long way off. Unfortunately, the policy of looking for quick returns cuts down the funds for research into this kind of reaction. The Labour Government axed British fusion research almost out of existence; the Minister of Technology at the time thought in his wisdom that there would not be an economical return in the 'foreseeable' future from this source of energy.

But no matter what kind of fuel we use, it will cause an increasing amount of environmental trouble. The heat lost into the environment at the beginning of the 1970s is already equal to 1/2500 of the total solar radiation (the energy that goes into heating the atmosphere and evaporating the water) received at the Earth's surface. Within the next hundred years this injection of energy could amount to as much as 1/20 of the solar radiation. By the end of the century heat loss into the atmosphere for the North American continent alone will be more than 190 million BTU's a year, and most of this will be released from the developing megalopolis of the eastern United States. And to keep up the energy production necessary for this one conurbation, even as near as 1980, nearly a third of all the water flowing in all the United States rivers and lakes will be needed for cooling in the power plants. In Europe similar conditions are developing.

Man himself is a source of heat; his own metabolism produces it. Multiply this by ten million, the population of London, and add it to the enormous quantities from machinery, lighting, vehicles, furnaces and all the sources that make a large city; multiply this by ten or even a hundred times for the megalopolises at the end of the century. Such massive injections of heat will cause widespread disturbances in the atmosphere and bring about local climatic changes. Already 'mountains' of warm air are encountered by aircraft flying over cities. Already —the heat problem apart—high buildings cause turbulence and, in addition to the stresses imposed upon the structures, funnel high winds along the streets, to the discomfort of the citizens. If, at the present rate of increase, the estimated 52,000 million world population could be sustained by 2120, the heat problem would be enormous; the huge upwelling of warm air from the major centres of population would alter the climate and disturb the air flow so much that the open country-side, if there were any left, would be subjected to continuous high winds making it impossible for anything exposed to live. Food production would have to be highly mechanized and under cover. Major changes in land, sea and air transport would be necessary, not only to move the massive population about, but to enable it to *move at all* under the new atmospheric regime. Cities are erupting like volcanoes; the land disappears beneath wave after wave of concrete. Architectural pollution has to be seen to be believed, homes like rabbit hutches cluster the roads and are valuable only if they have another little hutch attached to them for a motor-car.

The motor-car has made most cities unbearable to live in. Apart from the continuous noise of its infernal combustion engine, it is responsible for the worst fraction of the atmospheric pollution. It fills the streets with poisonous carbon monoxide and tetraethyl lead, which might be responsible for the sluggishness and frayed tempers of the citizens of London, Los Angeles, New York and Tokyo. Many main city thoroughfares have almost lethal concentrations of toxic gases during the hours of greatest human activity. At traffic lights when engines are idling the production of carbon monoxide goes up and the alertness of the drivers goes down. In the next decade, if the number of vehicles continues to increase, the carbon monoxide levels might be such that many streets will be too dangerous for pedestrians to use during the day, particularly in summer months—the streets will become canyons of death. But instead of banning the motor-car from the cities the trend has been to make the city more accommodating to it. Millions of pounds and hundreds of hours have been swallowed up in devising flow systems for traffic, but comparatively little money and time have been devoted to eliminating the gases, noise and the physical and mental strain imposed by the motor-car. The priority accorded to the motor vehicle was demonstrated in 1970 with the opening of the £30 million Westway extension of London's Western Avenue. At this point the elevated road, which took four years to build, is on average thirty feet from the rows of terraced houses where the residents were already subjected to the scream of jet airliners passing overhead and the rumble of underground trains below. The road was built to carry thousands of vehicles a day through some of the most overcrowded housing in London. No wonder the residents carried banners with the slogan 'Get us out of this hell!' The planners ignored people!

It does not seem to occur to governments or councils or their planners that the transformation of a city into a complex of motorways might prevent it from fulfilling its essential function as an environment for people. Los Angeles, where an estimated 60 per cent of the land surface consists of roads (compared with an average of 25 per cent in Europe) is the city *par excellence* for the motor-car. There are drive-in cinemas, drive-in banks, drive-in shops. It has become a city of which someone said that, if a pedestrian were rash enough to venture into the streets, he would be arrested as a vagrant.

In Great Britain we are already overcrowded; with nearly twelve million vehicles on our roads, we are beginning to produce conditions which few of us can tolerate. In thirty years' time there will be nearly forty million—that is, if we have the raw materials to build and the money to buy them and the roads on which to drive them. The conditions then will be nightmarish. We will suffocate in the pollution and be deafened by the noise.

Noise is one of the perils of urban life. Apart from motor vehicles there is the noise of aeroplanes, railway trains, factory machinery, typewriters and incessant talk to assault our ears, ears that were designed to pick up the rustle of leaves or the crackle of a twig in a forest glade. It is no wonder that urban Man has poorer hearing than the country-dweller! The damage done by noise can be alarming; of course it varies from individual to individual, but generally prolonged exposure to noise levels greater than 85 decibels (the decibel is the unit of sound) may damage the average man's ear permanently. The louder the noise and the longer the individual is exposed to it, the greater the risk of permanent loss of hearing. And noises above this level are common, especially in factories and near to airports.

1 ‘Today, when land itself is a precious natural resource, we are only now beginning to realize how much of it has been utterly ruined.’

2 ‘The countryside is spoilt at every turn by the leavings from John Bull's excursions into it. . . .’

◀ 3 Many have died, many more will die through the enormous injections of radioactive materials into the atmosphere from atomic and hydrogen bomb tests.

4 'The tall factory chimneys belched their foul fumes into the atmosphere. . . .'

5 Rubbish burning contributes to New York's serious air pollution problem.

6 The beach at Rovigliano (Naples), near the estuary of the river Sarno.

Young people who frequent discothèques, where sounds which are on the threshold of producing actual pain are usual (120 decibels), are quite likely to have their hearing reduced to that of a person of about sixty and will eventually go deaf.

Airports are among the noisiest places on Earth and when they are near cities they place intolerable stresses on those who live around them or under the low-level flying lanes. Aircraft landing at London's Heathrow fly over one of the biggest cities in the world and approach it over one of the most densely populated areas of the English countryside. If the Vertical Take Off and Landing aircraft (VTOL) were to be introduced for city-to-city transport the problem would be insufferable, because the engine noise of the VTOL aircraft is greater than that of conventional aircraft. Before any decisions to go ahead with this form of inter-city transport are made, this problem must be solved.

Supersonic aircraft such as Concorde present a new problem, that of the sonic boom. To a limited number of people living near military or experimental aircraft establishments it is already a problem. The sonic boom sweeps over the countryside in a wide carpet following the aircraft as it passes over. We can expect that if supersonic civil transport is accepted, most of the United States, the United Kingdom and the rest of Europe will be subject to its harrowing effects, for although most of the countries involved have banned flying at supersonic speeds over their territory, as its economic benefits begin to show, the will to keep the land 'clean' may be ever so slightly eroded. Once we had a ban on coal burning!

The inhabitants of the city are dependent on the efficient working of its machinery. The daily ebb and flow of commuters rely on the availability of fuel and power, and a stoppage of the supply or a break in the controls can throw the whole complex into chaos. In 1964 New York and other cities in the eastern United States were plunged into darkness through a power failure. People were trapped in subway trains, stuck in lifts and stranded in the streets for hours; the fabric of American civilization was torn from end to end by this sudden breakdown of the machine. As cities get bigger so will their requirements increase and greater reliance will have to be placed on automation, which in turn will need more power and control systems; but we cannot guarantee even now that power will always be available or that the controls will always function. So, the greater the degree of specialization and the greater the dependence on the city, the lower will be the ability of Man to survive. We shall have become like hot-house plants: without careful nurture we shall perish—a situation forecast many years ago in E. M. Forster's short story 'When the Machine Stops'. Our concrete and glass 'caves' may be more comfortable to live and work in, but they do not offer the stimulus that most of us need if we are to function well as individuals. Out of doors we should still have to face the untamed and probably untamable environment; only by completely encasing our cities should we ever be protected from it. The Man of the city today, a pasty-faced copy of his more robust ancestors, is fighting to survive in an environment he has created but to which many people are unable to adapt. And there is little evidence that those who seem to be adapting would be able to overcome the problems inherent in the large conurbations of the future. In the city, problems, like people, are tightly packed together; the relationship between pollution and population is more apparent. The relationship between resources and waste is not an abstraction, it is measured by the index of the cost of living. The conflict between man and man has reached such serious proportions that, unless answers are found very soon, the social diseases that began with the foundations of the industrial towns, and have been chronic since the turn of the century, will have reached an acute stage. The city is our greatest creation and may be our last.

CHAPTER 5
The garden of errors

Nowhere are the ravages of industry and the 'pollute and be damned' attitude toward the environment more evident than in our treatment of the land. Mining, quarrying and other extractive operations have left huge scars and spoil-heaps. Ugly urban spread continues to swallow up good land and the population load puts intensive pressures on the farms.

No matter what aspect of the pollution problem we consider, it is the pressure of population that dictates its size, complexity and ramifications. When Man was thin upon the ground, the environment could take care of his wastes through the natural processes, but as he increased in numbers so nature's capacity for coping with his wastes decreased. Not only has the *amount* of waste increased, but the nature of the products has become more complex. We are rapidly reaching a position today where the capability of the Earth to absorb waste is reaching its limits.

More and more land is gobbled up to meet the demands of the expanding population; space has to be found for houses, factories, schools, hospitals, offices, universities, roads, railways, docks, airports, reservoirs and power stations. In the United Kingdom urban sprawl is devouring the land at a rate of some fifty thousand acres a year, and by the year 2000 two million acres more will have been swamped. But while such development meets the housing and material needs, it negates its own advance by using up land which would otherwise have been valuable to agriculture; it is no use living in a nice house surrounded by material goodies if you are starving. How, there-

27

fore, is an increasing population to be fed on a decreasing acreage of agricultural land? This is one of the greatest problems facing highly developed countries, in particular the small ones like Britain with a high population density, for with even higher rates of population growth elsewhere, such countries cannot continue to rely on imported foodstuffs to make up their requirements *ad infinitum*, even if they could afford them.

Allowing for the decreasing acreage available to agriculture, an 80 per cent improvement in agricultural production would be needed to feed the estimated population of the United Kingdom by the end of the century. The only way of increasing food production is by more intensive farming, which brings its own problems. Already the developed countries have abandoned their traditional farming methods: farmers have had to resort increasingly to technological help and a vast armoury of chemicals, chemicals that are not found in nature and that consequently nature cannot cope with.

In growing food crops the farmer alters the natural ecology and introduces a new kind of habitat, one in which a single type of plant is grown. This monoculture not only provides us with food, but also offers an inducement to those insects and other creatures that also feed on these food plants, to reproduce madly in response to this free bean feast. For example, in part of south-eastern Russia scientists found that there were 312 different species of insect living in the area before cultivation; when the land was farmed, the number of species was reduced to 135, but the density of the insect population doubled. Some species increased as much as twenty-fold, and one, the wheat thrip, was 360 times more numerous than it had been under the wild conditions. Unchecked, these insects would swarm down upon the crops and rapidly devour them. It is at this stage that an insect becomes a pest. It is as well to remember that the vast majority of insects are quite indifferent to Man and his wonders and some are even beneficial to him. However, under the altered regime some of them compete directly with him for the food, and he counters them by using pesticides.

Unfortunately insects have an enormous capacity for reproduction, and in a few generations those able to survive, due to
natural immunity to the chemicals hurled at them, produce
offspring which can withstand what was previously a successful
pesticide. The farmer then tries another weapon from his
arsenal, and so the fight goes on. At each stage the chemicals
being used become more complex and alien to nature. Unfortunately pesticides, in general, do not destroy the pest insects
alone, they also destroy many benign and useful ones, including
those that prey upon the pest, so that inadvertently we are
destroying the very organisms which under natural conditions
help to keep the pest in check, and since these are nearer the
top of the ecological pyramid, they suffer more and their chances
of finding immunity are fewer.

So as the struggle continues a situation develops in which the
selectivity brought about by modern agricultural methods
produces a system consisting of three components: man, food
crop and pest. Rather than producing a new system in which
there is a co-operative regime, we have produced one which is
totally dependent upon the chemist, and one which at any
time can break down; the probabilities are ultimately in favour
of the insect. It is, therefore, a false premise to regard pesticides
as totally beneficial, but for the present it seems we have little
alternative but to use them in spite of hidden dangers. The
same reasoning is applicable to herbicides and fungicides; in
fact it could be argued that in these cases the competition is
even more favourable to the pests because their reproductive
rate is even greater. A point to remember is that most crops are
the product of generations of selective breeding and protection
by man, and are, therefore, weaklings compared with their more
robust wild competitors.

Accepting that without pesticides, herbicides and fungicides
it would be impossible to sustain the present level of food production, it is vitally necessary to understand their action in the
environment, to use them rationally and to control their distribution. The overriding consideration is that they are poisons;

to fulfil their function they are designed to be poisonous, and from a biochemist's point of view, the 'cabbage is brother to the king'. The similarities in living organisms are striking, not the differences. If a substance can destroy life at one level, it can be presumed to be at least a potential danger to life at another. For instance, if a chemical is found to kill rats at a certain concentration in the tissues, then it is likely that it will kill other mammals, including Man, at some level as yet unspecified. Another property of pesticides which makes them markedly dangerous is that they are persistent: they have been found in soils years after their application as nature cannot degrade them. DDT (2, 2-bis(p-chlorophenyl)-1, 1, 1-trichloroethane), for example, which has probably been the most widely used and publicized of all pesticides, is a particularly persistent chemical; in the environment it eventually breaks down to an even more persistent chemical, DDE.

Anything that enters the energy cycle and cannot be degraded is a potential danger to the system, and to understand what happens and to be able to recognize any damage we should look closely at the system step by step. In the countryside the trees, plants and animals are parts of a very complicated energy cycle which in its simplest form can be expressed as a chemical system dependent upon sunlight for its primary energy. The simple green leaf of a plant is a factory in which many of the complex chemicals essential for life are made with the aid of sunlight. So first we must ensure that the plants get sufficient sunlight, carbon dioxide from the air and nutrients from the soil. It is sobering to think that mankind is dependent on a very thin wafer of soil only a few inches thick. Of all the land masses, which in any case only cover about a quarter of the globe, less than half is usable by man. The world is a small island indeed and yet we have a continuing story of destruction of fertile soils through misuse. As we saw in the first chapter, most of the deserts of the world, and many of the barren islands of the eastern Mediterranean, are the result of intensive grazing and crop production, and in more recent times large areas of the U.S.A.

have suffered similar fates. Without artificial fertilizers the same would be happening in many more countries today. But the presence of pesticides and herbicides in the soil may negate any benefits from fertilizers in both the short and long term. The soil, thin as it is, is a living community with its own fauna of protozoa, earthworms, mites, insects, spiders and the like, all interdependent and all based on the cycle of energy within those few inches of humus. Dead leaves and other plant material (and animals) fall onto the surface and are broken down by micro-organisms into simple inorganic chemicals. Some liberate nitrogen into the air while others 'fix' it into nitrogenous compounds which are used by the growing plants to form the complex materials of life. In this cycle the earthworms play a vital part. Their movements through the soil keep it well aerated and drained, which allows the micro-organisms the right conditions for their work. The worms also bring down the leaves from the surface; these form the humus. Thus the worms keep going a continual process of enrichment. Earthworms are destroyed by many pesticides, and soils where the earthworms have been eradicated soon become stale and die and need an enormous amount of attention and money spent on them to keep them productive. If they do not have this, they either become water-logged or lose their plant cover and blow away in the drying winds. In either case they are lost to agriculture and amenity.

The first thing therefore is to ensure that the use of pesticides and other chemicals does not destroy the regime of the soil. It is just as important to protect the small birds and mammals that live on the soil fauna because in turn they act as the biological checks on the larger members, and without them there would be a breakdown through over-population of one or other of the soil animals. In turn, the larger predatory animals such as foxes, badgers and birds of prey keep the smaller birds and mammals in check. Because of their position at the top of the food chain the larger creatures are particularly vulnerable to the effects of poisons introduced at the base of the chain, for at each stage of it the poison is concentrated—the larger the organism, the greater the number of smaller organisms needed to satisfy its food requirements. The same principle applies in the freshwater and marine life systems, as we shall see later. Thus, if there is anything wrong in the food chain, we can expect the creatures at the top to show the first signs. Twenty years ago the peregrine falcon, kestrel and golden eagle populations in Europe suddenly declined, as did those of the peregrine, bald eagle, osprey and Cooper's Hawk in the United States. The regions where this happened coincided with areas where DDT and dieldrin, another member of the chlorinated hydrocarbon family, were widely applied. It was proved that these chemicals had poisoned the birds, also making their eggs sterile and their shells much thinner, resulting in much reduced breeding.

Over the years there have been increasing incidences of this kind, but until comparatively recently nothing much was done. Then suddenly we were rudely awakened to the possible implications they might hold for our own safety. Cosseted in our artificially bolstered communities, we tend to forget that we are ourselves dependent on the natural energy systems for food, water and oxygen. Man is at the very top of the food chain and if the birds of prey contain pesticide residues, then it is more than likely that he does too. The poisoning of these birds intensified research into the possible effects of pesticides, particularly DDT, on Man.

If we are to look for injury—that is, direct injury—to Man from modern pesticides then there are two obvious classes of people in the community in which we should be able to find evidence of harm: those actually involved with the making of the chemicals, and those, the agricultural workers, using them. In the first category there is little evidence of direct poisoning, but this might be due to the extremely good safety conditions under which they work. Many chemical processes in any case are carried out automatically and in areas to which the human operatives have little or no access. But even among agricultural workers the record, particularly in the United Kingdom, is extremely good, for although an average of one hundred farm-workers die through accidents every year, none of these (as far as we know up till now) has been caused directly by pesticides. On the other hand, in California about two workers a year on average die as a result of pesticides; but this may be because their use in the United States is more widespread and intensive than it is in the United Kingdom. Compared with the loss of life from accidents the risk from pesticides looks small, but it is a fact that people do absorb them, and in the United Kingdom everyone carries some two to three parts per million of DDT in his body tissues. The majority of us do not appear to be harmed, but the long-term effects could be different. There is also the possibility that damage from these substances is masked by secondary conditions which may cause the actual death and may themselves have been triggered off by the pesticides. The chlorinated hydrocarbons are known to alter the glucose mechanism and inhibit the enzyme adenosine triphosphatase which plays an important role in the energy economy of the human body. Work at the National Cancer Institute in the United States has shown that a concentration of only 46 milligrammes of DDT per kilogramme of body weight can produce a fourfold increase in tumours of the liver, lungs and lymphoid organs of animals, and some human cancer victims have been found to have two to two and a half times more DDT in their fat than occurs in the average person.

The increasing concern over the effects of pesticides has at last resulted in restrictions on their use. DDT and others, including aldrin, dieldrin and endrin, have been prohibited in the United States, and Sweden has declared a two-year moratorium on the use of DDT, aldrin and dieldrin. Canada has announced restrictions on the chlorinated hydrocarbon pesticides that are expected to reduce their use by 90 per cent; the United Kingdom, Holland and other Scandinavian countries all have tight controls. It could be argued that the benefits to food production that all the pesticides have brought and the outstanding success of DDT in reducing tropical diseases, with a subsequent saving of millions of lives, have more than compensated for their deleterious effects. But although justifiable on these grounds, there is absolutely no justification for indiscriminate use. The deaths of the wild mammals are a sufficient reminder that pesticides should be used with great care.

But once the wildlife stops dying and the directly observable evidence of damage is removed, it will be very easy (human nature being as it is) to slip back into the old system of indiscriminate pesticide use, for it will then be the pest that attracts attention. Thus it is more than likely that the accumulation in our fat will start again, and at some time in the future there might be an outbreak of unexplained diseases culminating in neurological disorders or death. The tragedy will be that if we find ourselves in such a situation there will be no way of ameliorating the condition; the chemicals will be there and cannot be got rid of by any known means which would not do as much harm to the body as the pesticides themselves. There would be no consolation then in saying that the necessary level of food production for the world had been made possible only by using these chemicals, and it is not an improbable situation. However, even taking the optimistic view that the controls will be maintained or tightened up further in the future, we are still not so sure that all is well—it may be that there is already sufficient of these chemicals in the environment to wreak eventual havoc. At the present time it may be the hawks that are sterile, tomorrow it might be Man!

The farmer must also keep down weeds or plant pests which, more often than not, are the indigenous plants of the area and will thus assert their superiority unless checked. Also, to make the maximum use of his land, the farmer may wish to clear scrub and trees. The chemical larder is raided once more to bring into action the necessary herbicides and defoliants. One group which is used widely is the 2, 4, 5-T (2, 4, 5-trichlorophenoxyacetic acid) family. These have been used for some years in Britain by the Forestry Commission, but their most widespread use has been in Vietnam, where the Americans latched on to the idea of ferreting out the Vietcong by clearing the undergrowth. Hundreds of square miles of the Vietnamese jungle have been laid bare by indiscriminate spraying, resulting in severe ecological damage. When 2, 4-D (2, 4-dichloro-phenoxyacetic acid) was introduced it was claimed that this was less persistent and of a lower level of toxicity to man, but two American zoologists, Drs Orians and Pfeiffer, who have made a special study of the effects of these chemicals, believe that it will take many years for the forest to become re-established. They feel that this might even be impossible except along the edges of river channels and backwaters, and furthermore the destruction of the habitat is equivalent to death for many animals. The damage has been wider than intended, for although the official view is that defoliant sprays do not drift outside the target areas, Orians and Pfeiffer have produced evidence that significant quantities of defoliants are regularly carried on the wind over much broader areas. Such an argument applies of course to any spraying; in 1970 there was a case in England, for example, where poisoning of fish in Chichester harbour was traced to the spraying of a farm near by with phosphate granules.

The chemical 2, 4, 5-T appears to affect not only the foliage and the soil. It was suspected of affecting unborn children when in 1969 many cases of severe teratogenecity (foetal deformation) were reported in a province of Vietnam which had been deluged with 2, 4, 5-T. Teratogenic effects were confirmed by experiments at the American National Cancer Institute. Their 2, 4, 5-T was also found to contain the impurity dioxin, and scientists at the National Institute of Environmental Health Sciences took the research further to see if it was the 2, 4, 5-T, or the impurity dioxin alone, or both together, which was teratogenic. Both the dioxin impurity and 2, 4, 5-T separately produced foetal abnormalities in mice, and as absolutely pure 2, 4, 5-T is not available commercially, the American Government thought it advisable, irrespective of which was the dangerous element, to ban the use of liquid forms of 2, 4, 5-T domestically and all forms on water areas. However, it is still legal to use the chemical for weed control on forest ground, pasture lands and footpaths. In Britain the Ministry of Agriculture has maintained that the teratogenic element is the dioxin which is present only in trace amounts in British supplies, but with the publication of the American findings, and backed by pressure from the forestry workers' unions, the Forestry Commission decided in April 1970 to suspend its use. So far, we do not know what the long-term effects of 2, 4, 5-T might be, although its chemical effect on plants and trees and on micro-organisms in the soil and its general toxicity are now well documented.

So far I have referred only briefly to chemical fertilizers, which must be used with the monoculture system of agriculture because under this system there is not time for the energy taken from the soil to be replaced naturally, and it has to be replaced artificially. The fertilizers, principally nitrates and phosphates, are designed to do this and are relied upon very heavily by farmers. On the whole they have done the job very well, but there are signs that there is a limit to their effectiveness. In Britain the Soil Association has been carrying out an experiment to test the long-term effects of their use: for some thirty years two very similar plots of land have been farmed in different ways, one with the use of commercial quantities of chemical fertilizers and sprays, the other by traditional methods and using organic fertilizers only. Striking differences between the two are now apparent; the traditionally farmed soil is friable and light, the chemically treated is now heavy clay.

The farmer, faced with diminishing areas of cultivable land, is inadvertently one of the greatest polluters of our environment. Some farmers are aware of the situation and try wherever they can to reduce the use of artificial materials, but there are a great many who refute that any of the chemical fertilizers, pesticides or herbicides are harmful. Their attitude is that the ends justify the means—the very essence of 'pollute and be damned'.

Over-grazing and intensive crop production are only a part of Man's abuse of the land. As soon as early Man realized the potential of wood as a fuel and as a building material, the world's forests were doomed. The depletion may have been negligible at first, but it was relentless, gradually quickening and then rapidly accelerating to the position today where vast areas of erstwhile forests are utterly denuded of vegetation (if of course they have not been swamped by cities and urban conurbations). Very early on the Chinese learnt the art of reaching and regulating intense heats in their pottery kilns, raiding the forests for their fuel. Today the legacy of a once thriving industry is barren hills and silt-laden rivers. The famous cedars of Lebanon have all but disappeared, victims of the demand for large imports of the wood by Solomon and, long before him, the Ancient Egyptians. Knowledge of iron-smelting hastened the denudation of forestlands. England, at one time densely wooded, had lost nearly all its forests by 1700, Scotland soon after. The United States, nearly all virgin land only two hundred years ago, will have used up its original timber by the end of this century, and until very recently substantially more saw timber was being felled than was being replaced by new growth. The tropical rain forests also have come under the axe, and even today, when we realize the importance of trees in the natural habitat, great areas of the Matto Grosso in Brazil are being felled. A tropical rain forest is essentially a natural photosynthetic factory and store of cellulose. In the ground beneath, sheltered from the Sun, a complicated but highly efficient process of converting organic material into the nutrients essential for new growth has developed. If the forest is felled, the floor is exposed to the intense heat of the tropical Sun and there is rapid oxidation in the soil, destroying the organic matter and organisms which make it productive and so draining its fertility. But trees are not the only source of energy, nor are they the only materials that can be fashioned into artifacts. Man learnt to extract fossil fuels and minerals from the Earth, and the soil then became a nuisance that lay between him and his new goals.

CHAPTER 6

The new wilderness

As Man 'advanced', he discovered that more and more useful substances could be extracted from the land, these in turn enabling him to get at others. Excavating, quarrying, mining, Man attacked with relish, but in his greed and thirst for 'progress' he gave little or no regard to what he left behind him. Today, when land itself is a precious natural resource, we are only now beginning to realize how much of it has been utterly ruined. Yet the dereliction goes on in the search for more and more materials to satisfy the needs of increasing populations, and for the quick profit. There are 250,000 acres of derelict land in Britain, 92,000 of which are buried 300 feet deep under the spoil tips of coal and china-clay mining; the Civic Trust estimates that the area is increasing by 3,500 acres every year. In addition there is an annual gross loss of land to mineral-working of 12,000 acres, of which coal extraction accounts for 6,000, and it has been estimated that over the next ten years 50,000 acres more will be given over to sand and gravel extraction. Much of this land is first-class farm land. In 1968 official figures gave the amount of derelict land as being 112,000 acres in England and Wales and 15,000 acres in Scotland. The discrepancy is accounted for by a difference in definition: the official definition of derelict land is 'land so damaged by industrial and other development that it is incapable of beneficial use without treatment'. This excludes tipping sites on which tipping has not been completed and any site which is still being worked, even though this may apply to only a small fraction of it. To the eye this land is just as derelict as abandoned workings, but this is a neat way for authority to persuade itself—and us— that the problem is not as bad as it really is.

Collieries are particularly repulsive, and an abandoned pit positively exudes decay—the lifeless tip, dominating the scene for miles around, the rusting buildings and machinery and the no man's land flanking the base of the tip, pock-marked with land-slips due to subsidence. Until comparatively recently no attempt at all was made to clear up the mess; the Coal Board has lost heavily on the deep-mining side of its activities and has carried out some reclamation only on the open-cast sites, which are more profitable. But even now colliery wastes are continuing to be tipped into the sea off the Durham coast, only to be washed back onto the beaches. Ten miles of the coastline are ruined, and as the collieries concerned are likely to be worked for another twenty years at least, there is a great risk that the devastation will go on.

Coal, clay, limestone, sand, sandstone, chalk, gravel, mineral ores, the land must yield them all. What matter the ugly spoil heaps sticking up like sore thumbs, the ravaged hillsides, the gaping holes frequently filled with murky water, the dust, the dirt, the stench? What matter that living in the midst of such dereliction saps the energy and vitality of the people? After all, why should they make the effort to keep their houses and gardens in trim when, after cleaning, the dust and dirt descend as thickly as ever? That is to be reckoned as the price of progress, reclamation and beautification are uneconomic and must be sacrificed. But we cannot afford *not* to reclaim the land. Every inch will soon be vital.

Some of the despoilers (but very few) have a conscience and try to do something about the mess their activities create. The Central Electricity Generating Board has a disposal problem of the same magnitude as that of colliery waste. It has to get rid of the pulverized fuel ash (PFA) residue from the coal-burning power stations, and one station alone can produce one million tons of the stuff every year. PFA is a good filling material and the CEGB is quite willing to bear the cost of transportation of the material to the required site. The Coal Board is not so forthcoming; it is rather ironic that both wastes arise from coal. However, in the United Kingdom the greater part of responsibility for land reclamation lies with the local administrative authorities. Some one hundred counties or county boroughs in Britain have derelict land within their boundaries and their reclamation record is lamentable, apart from a very few enlightened councils. A survey in 1968 showed that no fewer than forty-two authorities had made no effort whatsoever. The derelict land problem is not evenly distributed throughout the country: the worst affected county, strange as it may seem, is Cornwall, and this is due mainly to the china-clay workings. Six counties—Cornwall, Durham, Lancashire, Staffordshire Northumberland and the West Riding of Yorkshire—between them contain 65 per cent of all the derelict land in England; in Wales, Glamorgan is the worst sufferer. At the present rate of clearance, only about 2,000 acres a year, it will take fifty years to clear the current 'official' backlog, let alone that which will accrue from the further exploitation which is inevitable. The excuse for laxity and non-action is the age-old standby, cost. Land reclamation is admittedly a very expensive business, and frequently it is the poorer authorities who are the worst afflicted with derelict land. The fragmentary system of local government does not help because it means that control is very often vested in small units, and only close co-operation between a number of these could possibly raise anything like the necessary funds. Co-operation is a word bandied about by many, but applied by few: and then often only when it is convenient. In some instances, however, it is impossible to set aside money from the rates, the essential services demanding nearly all. But where money could be made available, the buck is all too often passed to the Government with the complaint that it does not provide sufficient grants.

Under the Local Government Act of 1966 all areas of Britain qualify for a 50 per cent government contribution to the cost of land reclamation schemes; small and poor authorities may under the 1966 Industrial Development Act also apply for an 85 per cent grant if the land is to be used for the development of industry. In some cases 95 per cent can be provided, although even then the authority cannot always raise the remaining 5 per cent. But the Government certainly does not push itself; local authorities are not required by law to put forward reclamation schemes, and even when they do they may be deterred by department fussing over minor details. Government departments can be simply downright stupid: a typical case was the M4 affair. The extension of the M4 motorway past Reading required 5 million tons of filling material and British Rail offered to supply part of it by transporting 2 million tons of colliery waste from South Wales and the West Country. This would have been cheaper than 'fill' dug up locally; it would have given the railway line a much needed boost, and saved much of the Berkshire countryside, an area of great natural beauty, from despoliation. One would have thought this idea would have been welcomed with open arms. But no, the fill had to come from local 'borrow' pits, which in turn have to be filled and restored with waste material from elsewhere. The Government is to provide 85 per cent of the cost to restore the countryside; if it had accepted the British Rail offer it would have avoided this and, more important, the destruction of the countryside.

Even though the Government makes grants for reclamation on the one hand, it can and does contribute to the problem itself on the other. The Ministry of Defence owns vast tracts of land which are given over to the demands of the military for the testing of weapons and for training its personnel to use them. These too in time become defunct, and the military leave the camp, depot or airfield. The land is sold, though frequently not straight away, and becomes the responsibility of the buyer. He more often than not does not do a damn thing to clear up the mess, but sits on the land and in turn waits for the next bidder, at of course the highest price. Or he simply sits.

Local authorities find that the most expensive part of reclamation is the acquisition of the land, and if such a problem is encountered at the start it needs a great deal of will and determination to go on. Apathy is the biggest obstacle to the solution of any pollution problem. No matter how hard a few enlightened people may fight, it is of no avail if they are baulked by lack of interest at every turn; it is very easy to give up. However, there have been some successes, notably in Durham and Lancashire where the local authorities have made a determined attack on ridding the landscape of dereliction. Lancashire has completed thirteen reclamation schemes, eight of which are in the North Makerfield area. The first to be tackled was an eleven-acre, sixty-foot-high flat-topped shale heap at the Bickerstaffe colliery. Today it is covered by 15,000 trees. In a scheme of entirely different character, a thirty-eight acre 'sink' of mining subsidence, chemical wastes, rubbish dumps and filthy water near Wigan has been filled in, levelled and grassed. Today it is a recreation area, one bright patch in a sea of grey. The biggest project, at Bryn Hall, was a 180-acre site of spoil tips and a huge 'flash' —a water-filled area of subsided land. A million tons of spoil had to be moved to produce a green pasture dotted with woods. Durham County Council has undertaken the levelling and landscaping of their colliery spoil heaps. The operation, which involves reducing the size of the tip and spreading it over a much wider area, and covering it with soil, has drastically improved the landscape. Ways have now been devised for seeding the tips directly with grass, thus cutting down the need for covering soil.

These examples show what can be done provided the will is there. Much more could be achieved if only the attitudes of the majority could be changed.

Suppose you have woken up to the devastation around you. What do you do? Blame the local authorities and the Government for sitting on their backsides? Blame the extractive industries for making the mess in the first place? But, think for a moment why the mess is being produced at all: to provide *you* with warmth, with housing and all the material things you like to surround yourselves with. And what about you? Are you ever satisfied? Don't you want more and more 'consumer goods' and new ones to replace the old? And what do you do with the things you no longer have use for? You throw them away. You add your own wastes to swell the tide.

The economics of the Western world in general and of the United States in particular are increasingly directed to the system of 'built-in obsolescence', the throw-away society. To survive, such a system must generate an ever-increasing demand whether people have any need for the goods it produces or not. The exploding American economy swallows up more than half of the natural resources won from this planet every year. People are deluged with advertisements urging them to buy, buy, buy; don't wait for a thing to wear out, you can't do without the new, improved model. Nothing escapes the treatment. Such a drain on natural resources just cannot go on. Already America is depending more than ever on foreign supplies: of seven metals most needed in making steel alloys only two are in adequate supply from domestic sources. Once an exporter of copper, America is now the world's largest importer; reserves of zinc and lead are so low that they will soon be uneconomic to work. Oil supplies are fast approaching depletion. Yet even when faced with the facts, the average American persuades himself that technology will solve all. Maybe it will, but it will be accompanied by high prices, shortages and a loss of personal liberty. America will not be able to rely indefinitely on imports from abroad; the exporting countries will soon need to retain their raw materials for themselves. The only hope lies in the recycling of waste.

To the waste from mineral extraction must be added that from agriculture, industry and domestic users. Agricultural waste is drawn from the excreta of animals, the slaughterhouse, crop residues, orchard and vineyard prunings, and the greenhouse. The modern system of intensive animal rearing produces vast quantities of manure. In the past this went back on the land as fertilizer, but today each farm generally produces so much that the soil just could not assimilate it and so it must be disposed of. Industry, grinding away to satisfy our material wants, churns out millions of tons of scrap metal, paper and paper-product wastes, slags, waste plastics and rags. And what about our direct contribution? Even though in terms of sheer quantity it may look small compared with the agricultural and mineral wastes, it is by no means insignificant and the very diversity of what we throw away makes up for the lower quantity—and it can be far more difficult to dispose of. In 1969 America produced over 4·3 *billion* (thousand million) tons of solid waste: 2,280 million tons from agriculture, 1,700 million tons from mineral extraction, 110 million tons from industry and 250 million tons directly from domestic sources. They threw away 30 million tons of paper and paper products, 4 million tons of plastics, 100 million tyres, 30,000 million bottles and 60,000 million cans and millions of tons of food wastes, vehicles and appliances, not to mention sewage—we will look at the latter in a later chapter. On average, Britons discard 19·5 million tons of rubbish a year and in 1980 this figure is expected to reach 23·5 million tons. The 1968 figure included 600 million plastic bottles, 6,000 million glass bottles and 10 million pairs of paper pants.

The throw-away society with a vengeance, the Americans are really going to town, steaks and other meats in disposable aluminium frying pans, muffins in throw-away baking tins, goulash in throw-away 'boil-in' bags, push-button omelets, fruit-whips, ketchups, whipped cream from aerosol throw-away cans and, wonder of wonders, the one-off, throw-away mouse-trap. This tribute to Man's ingenuity encloses the trap in an aluminium housing which snaps shut after the mouse has been enticed in. Throw away the lot, you don't even have to look at the mouse! In America packaging is a 25-billion-dollar industry. Each family spends on average 500 dollars on it every year, money down the drain in the literal sense. The British packaging industry is worth £800 million a year and is still growing. Each of us spends an average of 30p per week on packaging, and yet the expenditure on domestic refuse disposal is only 3p per person per week, of which 2½p is on collection and only ½p on disposal.

It is hardly surprising that the packaging boom has brought with it a change in the nature of domestic refuse. In the earlier part of this century most of the rubbish consisted of cinders and fuel ash, food residues, metal and glass, producing a dense and fairly easy-to-compact load. Nowadays the cinders and ash component is very small in comparison, its place being taken by more metal and glass—the one-trip bottle—and the new material, plastic. The mass is much more bulky and far more difficult to handle and compact. How is our rubbish disposed of? As far as most of us are concerned, refuse disposal is not a 'nice' subject to talk about, and once the dustman has carted our rubbish away it is completely forgotten—out of sight, out of mind. But let there be a strike, and it quickly brings home to us just how much rubbish we produce and how important is its removal.

In Britain the local authorities bear the burden of the disposal of domestic and trade refuse, while contractors cope with the industrial wastes. There are several possible ways of disposal, but whatever the method, all or part of the refuse ends up in some form or another on the land. The simplest way is to dump the refuse just as it comes, a stinking, insect-ridden, rat-infested mass, a hazard to health, pollution at its very worst. In these so-called enlightened days some local authorities still do this, but fortunately only a few. In the United States a very significant proportion of their rubbish is disposed of in this way.

A second method, and this is the main one in this country, is controlled tipping. In this method each layer of refuse must be covered on top and on the sides with at least nine inches of earth, ash or similar material. When tipping is completed the final layer should be covered with a few inches of soil and grass seed. But even this method has its problems: wind-borne litter is common, and bulky bits of refuse often protrude from the sides. There may be a lack of covering material, although this can be overcome by providing enough money to enable material to be brought in from elsewhere. But finding money for all the services is a difficulty to many local authorities, and refuse disposal, like land reclamation, very often goes to the bottom of the list of priorities. The problem of litter from tips is a difficult one to overcome, and controlled tipping can be completely acceptable only in deep pits or where there is natural protection from the wind. In such cases refuse can be valuable in filling in holes left by the mineral extractors. Yet much rubbish is dumped on good ground because the cost of transporting it to derelict sites is claimed to be prohibitive. A landowner near a town can make a small fortune by leasing a site to the local council for tipping. More often than not he does not live there and have to look at the operation day in and day out. This uncontrolled or semi-controlled dumping is rapidly becoming a national scandal in Britain. Apart from the obvious dangers inherent in rubbish dumps they are a threat to anybody who goes near them, particularly children. People often scavenge on rubbish dumps—they take anything, even discarded breakfast cereals. Recently a man salvaged what he thought was a drum of glue, but luckily for him he was intercepted and the local council had the drum's contents analysed. The result showed that if it had been ignited it would have produced the lethal gas, phosgene. This sort of dumping has to be stopped; it is a menace to the community.

Some 900 local authorities in Britain use the controlled tipping method, but 400 more use the semi-controlled method, a sort of half-way house. Many of these tips exceed the standard depth of six feet and are frequently uncovered. Levelling is done by a bulldozer and the exposed rubbish may be sprayed with insecticide. But this brings with it a very great risk of polluting water supplies from its possible seepage into the ground-water. We shall come back to this in a later chapter. This applies of course to any toxic wastes dumped indiscriminately on the land.

Pulverization is another method of refuse disposal. While this reduces the physical size of the mass, the bio-chemical processes still operate, similar to those of refuse in a controlled tip, but in this case decomposition is much quicker. Compacting takes the pulverization technique a stage further since it includes a mechanical fermentation process which produces a much more stable material. This can be used on some types of soil as fertilizer and would appear to be a very good example of recycling, but it has not been very successful so far. The material may contain toxic trace elements, it is costly to transport, it can be applied only at certain times of the year so that there is a storage problem, and the farmers have their own manure disposal problems. Only 1 per cent of Britain's refuse is composted and the technique has failed in the United States.

A significant proportion of our rubbish is incinerated, particularly in urban areas, but this by no means disposes of the lot. There is still some per 25 cent left: ashes, glass, metals and other non-combustible substances, and these all have to be dumped. Plastics are a particular menace—and remember, their use is spiralling upwards. They don't degrade under natural conditions, and although scientists are seriously researching into ways of making plastics which would degrade, they have a very long way to go yet. If burnt, plastics are notorious for fouling up incinerators, and PVC (polyvinylchloride), which produces hydrochloric acid on burning, has been blamed for severe air pollution in Germany.

Whatever the method used, the fact is that local authorities tend to be far too lax about their treatment of rubbish and there is a severe lack of co-operation at all levels. Often several local authorities may be competing for one site, any benefits from research tend not to be passed on, and there is no co-ordination between the sources of refuse and the extraction industries who could use it as filling material. For all wastes the cost of transporting them to suitable sites is claimed to be prohibitive, but the time will come when this will have to be done. Until then, hideous eyesores of such workings as the Bedfordshire brickfields will remain to violate the landscape.

So much for the dustbin! If we put all our rubbish in it, that would be at least a step in the right direction at the source, but we don't. Ours must be one of the most litter-strewn countries in the world. The countryside is spoilt at every turn by the leavings from John Bull's excursions into it—paper bags, cigarette packets, sweet wrappers, bottles, cars, orange peel, banana skins—it is all there. Man has discovered the motor-car in which he can probe right into the depths of the country. Very few places have managed to escape the relentless advance. Think, next time you go for a picnic and leave your mark, you might want to go there again. Do you treat your own houses in the same way? And do you stop at the picnic? Have you not wondered what to do with your old furniture, your old washing machine or refrigerator—your old car? All too many people dump them in the countryside and forget about them, much the easiest way. Even if you decide to dispose of your bulky articles legitimately, do you know what to do? Under section 18 of the Civic Amenities Act 1967, local authorities are required to provide places to which such refuse may be taken without charge by people in the area, but they cannot be prosecuted for failing to do this and many of them do not. This Act also imposes penalties of £100 on first conviction and £200 or imprisonment (or both) for a subsequent offence on any person who, without authority, '(a) abandons on any land in the open air, a motor vehicle or any part thereof, or (b) abandons on such land anything other than a motor vehicle being a thing which he has brought to the land for the purpose of abandoning it there'. Small price to pay, even if that person could be caught in the act! In 1972 in Britain some 600,000 cars will be scrapped and a not insignificant proportion of them will be unobtrusively removed to the countryside and abandoned.

Motor vehicles are one of the greatest polluters of our environment today both in themselves by their sheer numbers, whether operational or not, and by their noise, and in the products of their function, the exhaust gases which form one of, if not *the* most serious of our air pollution problems, as we shall see in another chapter. But their visible pollution is that on the land, the roads scouring the countryside, thronged with millions of speeding bugs, the rusting, rotting derelict hulks, the all-powerful all-pervading motor-car.

At one time many of the constituents of refuse were reclaimed, but nowadays salvage is uncommon. The economic system is geared to the extraction of raw materials from virgin resources, and vast sums have been expended on the technology required. The processes for recovering materials from waste are far less sophisticated. Theoretically nearly 85 per cent of all the paper and paper products thrown away is recoverable, but in Britain we only recover 30 per cent of it; the Americans recycle 19 per cent, and 15 per cent is retained or transformed in manufacturing processes. The Japanese manage to recycle 50 per cent of their waste paper, the level we attained during the Second World War. The Coal Board's tips contain 200,000 million tons of slag, most of which is reusable shale; yet only 6 million tons annually are put back into use, mostly for road foundations. The china-clay spoil-heaps could yield 150 million tons of sand suitable for concrete making. Three to four million pounds' worth of recoverable copper and nickel find their way into tips and sewers, and power station soot contains valuable vanadium. Five million pounds' worth of sulphur disappears in sulphur dioxide which creates its own severe pollution problem, as will be described in a later chapter. At a time when there is already growing concern at the depletion of the world's natural resources it is worth taking a very close look at what we throw away.

CHAPTER 7

Quiet flows the effluent

After the land the most obviously polluted parts of the environment are the rivers and lakes. We have lived so long with filthy rivers that we have ceased to notice that they *are* filthy. Suddenly we have awakened to the horrible fact that most of them are no more than sewers. We are now beginning to pay the price of neglect.

A fresh-water famine threatens the world. It may seem unbelievable that water could ever be scarce on a planet that is more than two-thirds covered by it, but the truth of the matter is that only 2 per cent of this huge amount of water is available for consumption—of the remaining 98 per cent, 97 per cent is salt water and 1 per cent is locked up in glaciers and ice-caps. It is not that the total amount is getting less—in fact it has remained more or less constant for as long as Man has lived on the planet—but the usable amount is dwindling because of misuse, waste and pollution. At the same time the ever-increasing world population is pushing the demand up and up and the water has to be shared by a greater number of people. An increasing demand for a diminishing resource can mean only one thing—trouble! It is ironic that this fresh-water shortage will have its greatest effect in the highly industrialized countries, the so-called 'have' countries of the world, for it is in these that more and more fresh water is needed not only for life itself but also for the industries which keep these nations prosperous. The needs of the United States will be doubled by 1980 and trebled by the end of the century. Britain's requirements will double by the end of the century, and taking the world as a whole, four times as much water will be needed then as at present. Already in many places the demand is beginning to exceed the supply. Taking the United States as an example again, Arizona uses just about twice as much water as it gets in rain. In Britain, especially in the south-east, Merseyside, the Midlands and the lowlands of Scotland, there could be many thirsty people by the end of the decade unless something is done.

Even if a crash programme of population control were to be introduced now, the situation, as it is, still means that the demand will increase; a situation we have to accept. But if we are to have any hope, we must make every effort to stop wasting and polluting our water supplies. This is the most urgent problem facing us today.

Unfortunately these forebodings will go largely unheeded because, except for those directly involved in supplying the water to us, most of us take water completely for granted. On a hot summer's day, just when there is the greatest need to conserve fresh water, thousands of us sprinkle our lawns or, even worse, wash our cars. Many thousands of us think nothing of leaving a tap running or dripping—it is too much trouble to buy or fit a new washer—and use far more water than we really need. Before water was available on tap, the average British domestic consumer used only about five gallons a day; now he uses between thirty-five and fifty gallons a day, of which four gallons are wasted because of dripping taps and leaking and burst pipes. How easy it is just to turn on a tap, and how cheap water is! At approximately $17\frac{1}{2}$p per 1,000 gallons for the domestic user and 5p per 1,000 gallons for the industrial user, the British bill is very small. In countries where water is scarce, Kuwait, for example, where 1,000 gallons costs £1.50 for the domestic user and 60p for the industrial user, people treat it with much more respect. Perhaps we in the West would too if we had to pay those prices. There is no doubt at all that we shall have to pay more for our fresh water in the not too distant future. Industry is already having to recycle most of the water it uses and that which we drink has been used several times over.

But apart from the sheer wastage in using far more water than we really need, we lose a colossal quantity through pollution. In 1958 the Ministry of Housing and Local Government undertook a survey of Britain's rivers which showed that a total of 1,278 miles were grossly polluted and 4,144 miles needed improvement; combined, these represent 27 per cent of the total length of the rivers in England and Wales, and the problem is acute because the worst pollution is found in the most populated areas. The Institution of Water Engineers in a 1970 report estimate that 'nearly as much as a quarter of the England and Wales water supply is drawn from sources that are frequently or continuously polluted in degrees varying from slight to dangerous, and a further quarter from sources liable to sudden pollution, varying from slight to severe'.

One of the worst problems concerns the River Trent and its tributaries, particularly the Tame, which drains Birmingham and other West Midland towns: these are so polluted that some people believe that they can never be cleaned. The St Austell and Luxulyan rivers in Cornwall have been written off by the River Authority, and the Douglas, a tributary of the Ribble in Lancashire, the Colne and Calder in Lancashire and Yorkshire, the rivers controlled by the Yorkshire Ouse and Hull Authority are all badly polluted. Up to some ten years ago the Thames was not much different, but its quality has gradually been improving through careful management by the Thames Conservancy. Some of the river estuaries are particularly bad since they are not subjected to the more rigorous controls which have been introduced in an effort to improve river quality. The Mersey is probably the worst polluted river, closely followed by the Tyne and the Humber.

The problem is not confined to Britain. The situation is the same or even worse abroad, and the description of many British rivers as 'open sewers' can equally be applied to rivers in Europe and the United States. The rivers of France, for example, empty some 6 billion cubic feet of water into the sea each year and of these 210,000 million cubic feet are polluted, or as someone once put it, the equivalent of 10,000 trains each weighing 600 tons! As the Seine flows through Paris it picks up a load of pathogens (disease-producing organisms): above the city the river contains about fifteen pathogens in each cubic centimetre of water, below it the number rises to over $1\frac{1}{2}$ million per cubic centimetre. The Rhine is so badly polluted that it is known colloquially as the sewer of Europe, a drain 530 miles long from Lake Constance in Switzerland to the Netherlands. When it passes through the valley of Grisons it contains fewer than 100 pathogens per cubic centimetre of water; by the time it runs into Lake Constance there are about 2,000 and below Kembus the number has increased to between 100,000 and 200,000. Although we know relatively little of what is happening in Russia, there have been reports of severe pollution of the Dniester and Volga rivers.

What is causing this pollution and how does it do it? To answer these questions we must understand the mechanism by which we get our water. Fresh water in the rivers and lakes is part of one of those beautifully balanced systems we have come to understand to make up what we call the environment. Our fresh water comes to us from the oceans via the atmosphere as rain, snow and hail. Part of the precipitation is evaporated back to the atmosphere by the heat of the Sun. What is left either runs off directly into the rivers or through the drainage systems of urban areas and down irrigation canals, or soaks into the soil. Some of the water percolating the soil is utilized by the plants and the rest seeps down to join the underground water system where it is called 'ground-water'. The plants take up the water required for their metabolism through their roots and the water produced by their internal activities is lost through the leaves in transpiration back to the atmosphere. As well as taking in water by drinking, animals use the water tied up in the plant cells when they eat vegetation. The water is released into the environment during respiration and excretion and when animals and plants die. It then evaporates or joins the ground-water, streams and rivers and returns ultimately to the oceans where the cycle begins once again. This is a continuous process, but its balance can be very easily upset by Man's interference. He has only to introduce substances foreign to the system or an excess of any of a number of substances already present to do untold damage. These are the 'pollutants'; leaving aside the atmosphere and the oceans, let us examine their effects in the fresh-water streams, rivers and lakes.

Far and away the biggest culprit in fresh-water pollution is sewage, domestic and farm, which if untreated makes tremendous demands on the oxygen dissolved in the water on which aquatic life depends. A survey carried out by the Water Pollution Research Laboratory on the Thames between 1950 and 1953 showed that 79 per cent of the pollutants in the river had their origins in sewage. A river with a high level of dissolved oxygen can tolerate a certain amount of untreated sewage, but only a certain amount. Fungi, bacteria and protozoa use this dissolved oxygen to oxidize the organic matter in sewage, which they use as food, and break it down into much simpler substances, carbon dioxide, water and ammonia; and there may be further oxidation to nitrite and nitrates. A convenient measure of the strength of sewage, and for that matter of water pollution as such, is the 'biochemical oxygen demand' (BOD), which is the amount of oxygen a standard quantity of sewage will absorb from water under standard conditions. The oxygen used up in this way is gradually replaced by slow absorption from the atmosphere but if, as is frequently the case, the discharge of sewage is continuous and excessive, the oxygen cannot be replaced fast enough for the system to cope. The organic matter is not fully broken down and the result is a foul-smelling, thoroughly objectionable stretch of water.

Yet despite this, millions of gallons of untreated sewage are still pumped into the world's rivers, estuaries and lakes, not to mention the seas, every day. In Britain, which is considered to have one of the best sewage treatment systems in the world, there is room for considerable improvement; three out of five of our sewage plants produce effluents below the River Authority standards. And it is here, on the subject of sewage, that the attitude of 'pollute and be damned' is particularly deep-rooted. Sewage disposal and treatment are the responsibility of the local authorities who have been, and still are in some cases, extremely loath to expend on it anything more than the minimum from the rates. The classic example is Edinburgh, which has no sewage treatment system at all but discharges 53 million gallons of filth into the Firth of Forth every day. In 1967 the city was told by the Lothian River Purification Board that a sewage treatment plant must be installed by 1973 at the latest. Nothing has yet been done; a trunk road, a police headquarters and an opera house are higher on its list of priorities! 'There are no votes in sewage' seems to be the common excuse for laxity.

I have already mentioned how badly the Tame is polluted, but this is hardly surprising when you consider that some 15 million gallons of untreated sewage are pumped into it daily from Birmingham and three other West Midland boroughs. In the words of the report of the Government Working Party on Sewage Disposal (1970), the Tame reveals 'ubiquitous shreds of sewage fungus and particles of miscellaneous obnoxious matter'. But to give it its due, the Upper Tame River Authority is now spending £17 million to staunch the flow. Huddersfield is another lax authority—of its 12 million gallons a day dry-weather flow of sewage, 2 millions are discharged untreated into the Colne, a mile upstream of its junction with the Calder. Above Huddersfield, this river still supports fish life but not for many miles below it. The Thames still has 95 million gallons of settled sewage pumped into it daily, but by 1974 another £20 million will have been spent on the system so that all the sewage effluent will be treated. In five years' time fish may be seen again in the middle reaches of the river.

Some rivers would not flow at all in dry weather if it were not for returned effluents. Half the water in the rivers Irwell, Tame, Rother, Mersey, Don and Warwickshire Avon is treated sewage effluent. All in all 3,000 million gallons of sewage are discharged every day in England and Wales and the volume is likely to double by the end of the century. Gradually more and more controls have been introduced to prevent discharge of untreated sewage into rivers, but, lamentably, the nearer the sea one goes, the weaker the legislation. The Clean Rivers (Estuaries and Tidal Waters) Act covers only discharges which began after it became law in 1960, and as many coastal and estuary towns had been dumping untreated sewage into the coastal waters for years previously, they can go on happily in the same way, and they do! I have mentioned Edinburgh as the classic example, but there are many more places: Margate, Scarborough, Weymouth, Bognor, Brighton, to name but a few. Can such resorts claim to be healthy?

The Report of the Working Party on Sewage Disposal stresses the point that purified sewage effluent is an essential part of our water resources and the discharge of treated sewage should be considered as a part of the water cycle. It further suggests that the administration of sewage disposal should be integrated with that of water conservation—it would be better to put it into the hands of the River Authorities. But, and most important, attitudes must change; sewage disposal must be seen as a 'vital industry rather than as a convenient means of forgetting about unpleasant wastes'.

Treating sewage to oxidize the organic matter is only half the problem. We have seen how the process produces nitrites and nitrates which pass through the filters into the rivers and lakes. The more the water is recycled the more the nitrates build up, and this is true of other mineral salts contained in the sewage, particularly the phosphates, which also pass through the system. To these must be added all the nitrates and phosphates in the run-off from agricultural land which has been doped with fertilizers and also from untreated farm sewage. In lakes and slow-moving rivers this can be very serious. A lake or river can normally support a rich variety of plant and animal life living together in another of Nature's delicately balanced systems, but if the system receives an extra charge of phosphates and nitrates, the green algae species thrive and multiply out of all proportion, resulting in a population imbalance. Millions of these small plants can, in a comparatively short time, completely cover the water, and by taking the oxygen out of it they leave very little for the other plants and the fish. The plants die, and the fish, if they cannot move to other areas, perish too. The end result is eutrophication, a sheet of stagnant water with a layer of algae outgrowing the capacity of the water to feed it. The dying algae ring the edges in their stinking millions. The water is dead. In North America the problem has been emphasized by the destruction of the Great Lakes. Millions of people depend on these lakes for their fresh-water supplies and at least 5 million are completely dependent on them for their livelihood. It is hard to believe that such vast tracts of water, covering some 95,000 square miles and draining 300,000 square miles, could become seriously polluted, but it has happened. Lake Erie is completely dead to aquatic life; Man has prematurely aged it by 15,000 years. Lakes Michigan and Superior, while certainly not as bad as Lake Erie, are gradually going in the same direction, and Chicago's water supply is already affected. Twenty zones on the American side of Lake Ontario also will soon be seriously polluted. Lake Zürich in Switzerland, and many other rivers and lakes throughout the world, are similarly affected. If nothing is done to arrest the destruction there can only be disaster ahead.

The levels of phosphate and nitrate above which the process of eutrophication can happen are 0·01 milligrammes per litre and 0·3 milligrammes respectively in the spring when the algae are developing. Are there ways of reducing these levels? In fact there are. Phosphates can be reduced with lime or ferric or aluminium hydroxide, nitrates can be treated with bacteria; 95 per cent can be removed in this way, but even so some enrichment would still occur. But the processes are expensive (about 1½p per cubic metre) and this has prevented their use. Another method is electrodialysis, which removes the mineral ions through a membrane under an electrical potential difference; and there is also reverse osmosis, which has a financial advantage over electrodialysis. These processes will gradually come into use, and it may be possible for the mineral concentrates they produce to be fed back to the land.

If the nitrate level rises above 20 parts per million water could become toxic, and even fatal to children under two years of age, a sobering thought considering that 15 parts per million have been recorded in several areas of Britain. But the nitrates are not the only dangerous substances originating from sewage and agricultural wastes. Over seventy viruses have been isolated in human faeces and all are found in sewage and pass through the treatment processes. Most of the infective hepatitis cases have been traced to virus-contaminated drinking water and the incidence of them is increasing.

But some of the most dangerous substances are the pesticides and herbicides which, as we have seen, have created havoc on the land. These, with the fertilizers, find their way into the rivers and streams. In some cases insecticides have been deliberately sprayed onto water to kill off insects and insect larvae, with sometimes disastrous effects. In water the action is similar to that on land; that is, the chemicals are concentrated in the life cycles of rivers and lakes. We have seen how land birds have been killed by pesticides, and this slaughter has been paralleled by deaths in water birds. A striking example occurred on Clear Lake in California in 1969 as the result of direct spraying onto water of the chemical DDD (2, 2-bis (p-chlorophenyl)-1, 1-dichloroethane), a close relative of DDT. It was sprayed to kill off the larval form of a midge which irritated the fishermen. Three applications were made in as many years. The substance was absorbed by the plankton in the lake, the plankton was then eaten by small fish in the large quantities necessary for them to live; thus the DDD began to be concentrated in the food chain. The small fish were eaten by large fish, which furthered the concentration of the chemical. At the top of the food chain the very large predatory fish and water birds, the western grebes, were feeding on fish with very high concentrations of DDD. Both died, the grebes being all but wiped out.

Lake Michigan has suffered particularly from DDT poisoning: 22,000 pounds of coho salmon, which had been introduced earlier to combat the sea lamprey that had got in the lakes, had to be destroyed on the orders of the American Food and Drugs Administration because of a high concentration of DDT residues. The dangers of pesticides to the Great Lakes were put forward strongly by nature lovers, but virulent denials were made by some officials who claimed that the threat had been exaggerated and that the needs of industrial progress and agricultural production were of paramount importance.

A very important characteristic of river pollution is that a pollutant or pollutants can be introduced into one part of the river and yet the effect may not be felt until farther downstream. As some of the world's most polluted rivers flow through several countries, incidents can occur to test the most cordial of international relations. Such an incident, which in fact turned into a major disaster, happened in the Rhine in 1969 when millions of fish died. The pollutant which caused this was traced to the insecticide 'Endosalran', of which only 220 lb. would have been sufficient to cause the damage which was done. In the case of this particular disaster, run-off from agricultural land was discounted as a possible source, and it was presumed that the insecticide had leaked from a barge—the West German research ship investigating the disaster pinpointed the origin to a part of the river where some barges had been seen. So strong was the poison that all the micro-organisms in the water were killed. Apart from the fish, which were virtually wiped out, water birds and mammals died in large numbers. When healthy fish were lowered into the river along the North Rhine–Westphalian section to see if they were affected, they died within minutes.

As the Dutch take about 60 per cent of their daily water supply from the Rhine it is essential that they get the earliest possible warning of anything happening upstream. In this case they had only a few hours. If the insecticide had found its way into the Dutch water supply there could have been a major tragedy with international implications. It was only the prompt action of the Dutch water˙authorities in switching onto reserves which prevented it. To be fair to the Germans, it was several days before anything unusual was noticed on the river, but then the dead fish were becoming more numerous as the poison spread until there were literally millions of them. Because of this time lag between the incident and its effects being noticed, the Germans were unable to give much advance warning to the Netherlands authorities; they hadn't much time themselves. But as soon as they knew, they launched an investigation. People were warned not to swim or wash in the river, or to eat fish or feed them to animals, and livestock were kept clear of its banks. The German scientists investigating the contamination concluded that it could take years for the river to recover sufficiently for fish to live in it again. In the end some 250 miles of the river were poisoned and an estimated 40 million fish perished.

One of the difficulties which faced the Germans, and indeed faces any country with similar problems, is that of isolating a particular pollutant in a river as badly polluted as the Rhine. There are literally hundreds of pollutants in the river which, had they been present in sufficient quantities, could also have destroyed the fish. Although sewage and agricultural wastes contribute the greatest *volume* of pollutants to a river, industrial wastes can increase their *numbers* quite startlingly. These too can draw on the oxygen reserves in the water.

Sawdust, wood fibre and fibrous textile waste which break down only slowly by bacterial action deoxygenate the water to such an extent that, apart from sewage fungus, plant growth is impossible. Rivers in northern England are particularly vulnerable to pollution by textile waste, the legacy of indiscriminate dumping going back to the heyday of the Industrial Revolution, very much a case of 'pollute and be damned'. Originally the culprits took their water from the unpolluted rich storage areas of the Pennines, but as water becomes more scarce the manufacturers are going to have to look at the rivers on their doorstep and will have to pay the penalty of a century of misuse.

Flue gases from some power stations are 'washed' to remove sulphur dioxide to reduce atmospheric pollution, but in water these react to form sulphites, which take up the dissolved oxygen in the rivers and form sulphates. In 1963 field tests on a power station showed that the effluents took up eight tons of dissolved oxygen from the river in the vicinity of the station every day.

As well as contributing their share of phosphates, synthetic detergent effluents produce foam which inhibits the reoxygenation of the river water. The newer types, the so-called 'soft' detergents (alkyl or aryl sulphonates), are an improvement because they are more readily attacked and destroyed by bacteria. Surface films of oil also hinder the reabsorption of oxygen, and if the film is thick the process can stop altogether. This is essentially a problem in slow moving water; in fast-flowing rivers the effect is generally too transient for it to have any lasting impression on aquatic life. In Lake Michigan, which particularly suffers pollution from industrial wastes, an accident at a detergent-producing plant led to 10,000 gallons of soya bean oil finding its way into the lake.

With so many different pollutants entering a river we must expect that they could react with one another to produce even more dangerous substances. A single pollutant may do relatively little damage, but in combination with others can produce active poisons. This happened in a river in Oklahoma. A textile mill continually discharged its waste into the river, but without any noticeable harm being done to the water; at least it was drinkable. Then a pesticide factory started operating and added its effluent to the river, but still the water did not seem to be noticeably affected. Later a bleaching company opened up and used the same river for its effluent. All three companies were careful about their effluents, but shortly after the bleaching company started its processes the water quality dropped and doctors reported an increase in digestive troubles among the local people. On analysis the water was shown to contain a poisonous chemical that had been produced in the river by the reaction of chemicals from the three effluents. In this case common sense prevailed and the three companies concerned changed their effluent treatment methods and everything was put right.

To sewage and agricultural and industrial wastes contaminating our fresh-water supplies must be added the seepage from rubbish dumps and spoil-heaps. This represents one of the most scandalous situations to be found anywhere. Many dumps are sited just outside towns and almost anything is allowed to be left on them, as we have seen in Chapter 6. In time, rain and water from melting snow leach out the polluting chemicals in solution. These enter the ground-water, and if the dumps are sited above a town they may flow into its own water supply. Polluting chemicals are also known to seep out of old mine-workings to affect rivers and water supplies many miles from the source of the contamination. In parts of Scotland seepage of floodwater from closed collieries is already causing concern because of this. The Don in the Pennines is polluted by the mineral ochre seeping from closed collieries in the Penistone area, but it is encouraging to see that the Humber and Ouse River Board are investigating a means of chemically neutralizing the ochre. The spoil heaps and workings of disused lead, zinc and copper mines in Central and North Wales pollute the nearby streams and rivers. And very slight amounts of lead and zinc (3 parts per 10 million) can kill most fish life. This pollution is probably an important factor in the high incidence of stomach cancer.

The combination of all these pollutants makes it very difficult to know what is actually in the water, and although standard tests are regularly carried out for the more obvious ones, many others can slip through. Even the most efficient sewage and water cleansing treatments cannot get rid of the invisible chlorinated hydrocarbons and many other persistent chemicals. They are in our drinking water whether we like it or not. The level of nutrient salts in our water is not harmful to adults as far as is known, but it can be poisonous to young babies. In addition to the nitrates there are other substances which may be harmful. In 1968, for example, random samples of drinking water taken by the municipal laboratory in the Liverpool University Department of Civil Engineering from eleven hospitals, four petrol filling stations, four police stations, three private houses, one public convenience and one university department, covering twenty-four towns and county boroughs and representing a population of some seven million people, contained 0·032 milligrammes per litre of phenolic substances, which is sixteen times greater than the World Health Organisation's maximum allowable limit. And 96 per cent of the British population are drinking water which is at times sufficiently acid to dissolve lead. Mercury is another dangerous pollutant which can get into the water supply.

Even though a substance can be present in quantities which may not in fact endanger health, the taste it imparts to the water can make it completely unacceptable for drinking. A few years ago a factory in the south-east of England was using phenolic materials for cleaning. The waste went through the firm's treatment plant, and this included a chlorination process which produced chlorophenols. The effluent went into the town's sewage works for further purification but the chlorophenols went through unchanged and were discharged with the rest of the effluent to the local river. Water was abstracted sixteen miles downstream for drinking and there was a rash of complaints. The objectionable taste was due to the persistent chlorophenol.

As well as being dependent on the oxygen content, acidity or alkalinity and availability of dissolved salts, chemical toxicity is also dependent on temperature, and temperature itself is a particularly important variable affecting the biological cycles in the water. An increase of as little as 4 deg. C. can speed up most of the biological processes, 5–10 deg. C. can have adverse physiological effects and can cause a redistribution of the flora and fauna communities. Cold-water fish such as trout, which incidentally have been used to monitor toxicity because they are extremely sensitive to impurities, die if the temperature goes above 25 deg. C. Carp can withstand 35–38 deg. C., pike a temperature between these two. Sewage fungus flourishes at high temperatures and the BOD increases. Thermal pollution can be a very real problem, particularly near power stations. A 2,000-megawatt power station requires 55 million gallons of cooling water every hour and this is usually obtained from a river or a lake. Even though as much as possible is condensed and recycled, a power station of this size still has to cycle 13 million gallons of river water an hour. The temperature of the river must rise. In stretches where deoxygenation is already severe and fish are at the limit of their tolerance, thermal pollution can tip the scales in the wrong direction and result in completely anaerobic conditions. Since power stations are usually sited where the river is polluted anyway, conditions are just right for this to happen.

Trouble from thermal pollution is foreseeable in the Rhine, for it is intended to build forty nuclear power stations along its banks. The French, West German and Dutch experts of the committee set up by the Rhine Anti-Pollution Commission to look into this have 'reviewed the situation', and left it at that; no government or international intervention is planned.

Yet another variable which can adversely affect the biological cycle is the presence of substances in suspension which cause turbidity and make the water opaque. The light intensity is thus reduced, inhibiting the growth of plant life on which the primary consumers live. The invertebrate animals cannot attach themselves to stones because of the layer of inert material which gathers on them. The heavier particles sink and smother the bed of the river, slowing the current, killing rooted plants and algae. Slimes and water-washed wastes are the culprits here.

With so much of the world's water supplies physically, chemically and biologically polluted, it is little wonder that the provision of fresh water looms large in any country's economy. More and more must be found and much of man's ingenuity has gone into the building of giant dams and irrigation schemes. Their importance is so great that they have become prestige symbols. If you wish to 'win friends and influence people' you offer to provide them with a major water scheme. The recipients are delighted; they can show off their new acquisition to their admiring but less 'fortunate' friends, and you have your influence! But such schemes can very easily backfire because they are so often considered, planned and executed with complete disregard for the biological and climatic changes their construction might cause. The wider implications are ignored. The formation of Lake Nasser in Egypt destroyed the Mediterranean sardine fishery which was worth $7 million a year to the Egyptians. The Aswan High Dam, by ending seasonal flooding and the deposition of silt, has led to the once fertile valley of the Lower Nile having to be treated with artificial fertilizers. Some of the power produced by the dam has to be used by plant specially built to produce the fertilizers. In addition, the water has brought a marked increase in bilharzia disease, or schistosomiasis, which has now replaced malaria as the most widespread disease.

The scale on which Man can alter the environment when he interferes with natural water regimes can be gauged from the effect that the damming and industrialization of the lower Volga have had on the Caspian Sea. This, the largest inland sea in the world, is being reduced in area at an enormous rate; already some of what were formerly fishing villages are a hundred miles or more from its shores and by 1980 its depth will have dropped another 8 feet. Technological considerations come before ecological considerations. What you gain on the roundabouts you lose on the swings—if the roundabout goes more quickly, so do the swings!

But if the roundabouts are technological advances then one of the swings is water pollution. We cannot afford to go on building more and more dams for our water supply in the hope that this will solve the problem. It won't. We shall have to rely more and more on recycled water and we must stop polluting what we have. Controls must be tightened so that no untreated sewage is allowed to be discharged into rivers, estuaries and coastal waters. Industrial waste must be treated at source and companies should be compelled to do this and also declare the composition of their effluents—their argument that money spent on this puts them at an unfair disadvantage in competition with others (particularly foreign firms) is no excuse. They could be out of business altogether before long through shortage of water. Penalties for pollution, at present negligible, must be stepped up and should cover accidental pollution. How many 'incidents' are called 'accidents' to avoid even the existing penalties? There are signs that the pendulum is beginning to swing in the right direction and that people are waking up to the situation, but every one of us must be made to realize that water is the most precious of our natural resources and should be treated as such.

CHAPTER 8

The familiar sea and other stories

All rivers ultimately empty their burdens into the sea. Those pollutants that have not been absorbed into the land and fresh-water life cycles find their way into the estuaries and eventually into the sea. I say 'eventually' deliberately, because the rubbish that comes down the rivers stays in the estuaries for considerable periods of time, as the tides tend to wash it backwards and for-wards. Normally an estuary is rich in wild life, which ranges from the micro-organisms in the sand or mud to the fish upon which sea-birds, marine mammals and generations of fisher-folk have lived—at least, until recently.

I was born and brought up in a town bordering an estuary and among my earliest memories are times spent collecting a wide variety of marine animals left in the pools as the tide receded. Only twenty years ago the shrimp and fishing boats still went out and their crews could manage to scratch a living from the shrimps, plaice, whiting and a few cod. Today that is all over, there are none of these—the Mersey estuary is com-pletely fouled. We could see the pollution in those days; human excreta floating down the river and out to sea, and back again on the incoming tide, was a common enough sight. The brown-ish lumps were known to us boys as 'Mersey Goldfish' and for all I know, the boys of New Brighton still call them by that name.

The Mersey estuary is perhaps the worst polluted of all the British estuaries, but it holds this position of eminence by only a small margin. Most of our big estuaries are fouled by sewage and domestic and industrial wastes. But pollution is not confined to estuarine waters—the British Isles are surrounded by a rim of liquid filth. There are very few places where you can swim today without being in danger of coming in contact with the physical evidence of sewage. One of the best comments on this state of affairs was made by Dr Reginald Bennett, M.P. for Gosport and Farnham, when he told the House of Commons that 'from many English beaches it is no longer possible to swim, all one can do is to go through the motions!'

Pollution of our coastal waters is severe and presents a serious health risk to the seaside residents and to the millions who flock to the beaches at holiday times and week-ends. Many of the seaside resorts that invite us with bright posters of scantily clad maidens on golden beaches to take a dip in their health-giving waters discharge their sewage, frequently untreated, into those very same health-giving waters—in fact no fewer than seventy-eight local authorities in Britain empty untreated sewage into the waters around our coasts. Nauseating as it is, faeces are in time broken down by bacteria and by mechanical action of the moving sea-water which breaks them up and allows sufficient oxygen to get at them for the biological processes to come into play, so that generally the worst excesses that occur in fresh water do not happen in the sea. But not all the contents of sewage are completely degradable. Sewers do not only empty out the products of metabolism, they also empty some 'one-trip' goods of the throw-away society—most of the world's annual production of 22,000 million rubberized contraceptives find their way into the sea by this route. These objects are not just unsightly, they are a culinary menace! Recently complaints of knives 'bouncing off', and of deformed neck regions of fish reached the ears of fishery scientists. They investigated the problem and found that what had been happening was that the rubber collars of the contraceptives floating in the water—for that is what is left after 'weathering'—attracted small fish which poked their heads through the rings, and if the rings got past the operculum (the gill covering) they stuck. As the fish grew, the rings became embedded. Hence the difficulty encountered by the would-be consumer in cutting his intended tasty morsel!

There is a loss of amenity and a very real health risk in the continuance of this policy of pouring untreated sewage into the margins of the sea, but more important is the threat that the pollution presents to the biological productivity of the sea. It is in the coastal waters and the shallow seas that cover the continental shelves that we find the main biological activity of the marine environment. The rivers bring down in suspension millions of tons of minerals every day and it is this which forms the nutrition of the phytoplankton—the plant plankton—the basis of all the food chains in the sea. It is because of this that coastal pollution is particularly dangerous, for as the pollutants come down the rivers they mix with the minerals and thus will be absorbed with the food into the life cycles of the littoral ecosystems.

There are four main types of seashore habitat: the muddy flats—usually in estuaries; the sandy and the pebbly beaches; and the rocky shores. In the mud flats there is a thriving community of bacteria, single-celled animals, worms and molluscs which provide the food of the familiar estuarine birds and a wide variety of fish. The sandy beaches have their own distinct communities, not as rich as the muddy shores but nevertheless important to the ecosystem, and even pebbly beaches, contrary to what some people have said, have very important animal communities. The rocky coasts with their shelves, caves and rock pools are the richest of all, with an abundance of life forms which have adapted to what are perhaps the hardest conditions of all in the marine environment. Each of these communities is important to the inshore ecosystem and vital to the coastal fisheries and oyster, shrimping, lobster and crab industries. All of them round the British coasts are rapidly becoming defiled. Musselburgh can no longer sell its mussels because they are unfit for human consumption; they have sat too long in the Forth estuary living on Edinburgh's untreated sewage. Mussels and many other molluscs feed by sucking water into themselves and from this they sift out particles of food. They are living filters, retaining and concentrating along with their food the persistent man-made chemicals and harmful organisms. And although neither the chemicals nor the organisms may affect the molluscs themselves, any predator that eats them is running the risk of harm and possibly even death. Just as in the fresh-water and land life cycles, dangerous substances are concentrated as they move up the food chain.

Coastal pollution is one reason why fish farming is never likely to reach the scale of development hoped for. Fish farming involves concentrating large numbers of fish in relatively con-fined spaces in shallow waters near to the coast, which means that around the British Isles and the other coasts of north-western Europe and of North America they would have to live continually in water with fairly high concentrations of infected sewage and industrial effluents. Fish hatcheries in particular need a very high standard of sea-water because many of the larval stages through which developing fish pass are extremely sensitive to the quality of the water. And although the United Kingdom has played an important part in developing fish-farming techniques, it will probably be the last place anyone would want to establish a commercial fish farm.

Sewage and disease-producing organisms are probably the lesser of the evils that bathe our shores. The greater danger comes from the industrial effluents and the agricultural chemicals, which are literally draining off the land daily in enormous quantities; all are being absorbed into the life cycles of the coastal ecosystems.

The effect of industrial pollution on marine life has been demonstrated very clearly by Dr David Bellamy and his team at Durham University. They discovered that in comparatively unpolluted waters of parts of north-eastern England, ninety-two species of seaweed are found, while only thirty-one species are able to live in the neighbouring polluted areas. This drop in plant growth indicates a reduction in the biological productivity, and Dr Bellamy's subsequent researches have shown that the destruction of the ecosystem in the coastal waters is speeding up, a situation which is extremely dangerous and must be stopped. It is sometimes argued that a comparatively local situation, such as that reported by Dr Bellamy, would not have any large-scale impact on marine life. But this argument fails when we examine the evidence of many more local situations, although we have to admit that marine scientists just do not know for certain what part the littoral communities play in the totality of the marine environment.

   The evidence is alarming because of the severity of the pollution and the subtlety with which the danger can creep up on us. Many of the most dangerous pollutants in the sea are invisible, and until 1969 were recorded only when some human tragedy occurred. Some startling discoveries have been made: for example, lead in the surface waters of the Pacific Ocean has increased from 0·02 to 0·07 microgrammes per kilogramme of seawater since the middle 1920s. The slow mixing of the water is keeping this deadly metal in the upper layers of the ocean where the biological productivity is highest. Mercury is another lethal metal which can remain undetected until it is well and truly ensconced in the ecosystem. In the middle 1950s a plastics factory in the town of Minamata on the island of Kyushu in Japan discharged large quantities of wastes containing mercury into Minamata Bay. The local fishermen, ignorant of what was going on, continued to catch their fish and to take their daily catch home to their families. Because of the contamination, 110 people, mostly belonging to the families of the fishermen, were killed or badly disabled between 1953 and 1960. At the height of the crisis in 1956, forty-two people were poisoned. Scientists examining the fish found that their tissues contained an average of 50 parts per million of mercury, but some individuals had as much as 102 parts per million. By 1958 the mercury had moved into the Minamata River itself, and in the following year people up the river were being poisoned too. The scientists traced these poisonings from the wastes of the plastics factory to the waters of the bay and river, to the fish and shell-fish living in the contaminated water and to the people who lived on the fish. But it didn't stop there. Some of the women who ate the fish passed the poison on to their unborn babies; some-times mothers who showed no symptoms of mercury poisoning themselves gave birth to babies with congenital defects.

Another disaster struck in 1965 and again it was in Japan and again it was caused through a discharge of mercury wastes. This time it hit Niigata on the Island of Hon Shu. In this case twenty-six people were poisoned, five of whom died. In both of these tragedies methyl-mercury was the pollutant. This is formed during the production of plastics when acetaldehyde is made from acetylene using mercury as a catalyst, and is one of several compounds discharged from plastics factories. The mercury problem is still with the Japanese. Every now and then high methyl-mercury concentrations are still found in aquatic organisms and quite recently fish from the Oyabe River were found to have as much as 4 parts per million. Mercury compounds have also been found in fish from coastal waters off the United States and Northern Europe, particularly the Baltic. It may be only a matter of time before the tragedy of Minamata Bay is repeated nearer home.

Mercury and lead are only two of the thousands of products, the biological effects of which are unknown, that are daily being pumped into the sea. One estimate gives the number of different substances reaching the sea at half a million. Among these are the ubiquitous organochlorine and organophosphorous com-pounds and many of them are now being found in marine animals. One group, the polychlorinated biphenyls which are used in a variety of industries—plastics, paints, lubricants, etc.—played an important part in the Irish Sea sea-bird disaster in the autumn of 1969.

In this incident some 10,000 sea-birds were destroyed (some estimates give a figure nearer 70,000), most of them guillemots, birds which with the puffins and razorbills form the family of auks. Some were apparently seen to be feeding on small fish off the Irish coast only hours, in some cases, before they were found dying. From then on things got confused, strong winds blew up, and afterwards thousands of birds were found dead and dying almost all round the coast of the northern Irish Sea and the Isle of Man. About half were covered in oil, but the others seemed, although in some cases a little thin, to be in a reasonable condition for that time of year when these birds are in moult.

A disaster of this magnitude could not pass without some action being taken, and the Natural Environment Research Council (NERC) speedily gathered together a group of experts to investigate the deaths. They included fishery and marine biologists, ornithologists, chemists and oceanographers. The facts as they were known were presented to the group and various theories to account for the disaster ranging from infection to pesticides were put forward. But by studying the timing of the various stages in the incident and relating this to the meteorological and oceanographical conditions, and bearing in mind what was known of the guillemots' habits in the Irish Sea, several people came to the conclusion that whatever had happened must have occurred in the area of the North Channel and after the birds had left the breeding colonies. This made it apparent that it was unlikely to be a virus, because once the birds are at sea they are well dispersed and any infection would not have become so widespread. Another idea was that while they were still at the nesting sites they had been dusted by a chemical which had had a delayed effect on them. It was also suggested that they had taken in DDT while feeding during the spring and summer and that this had been stored in their fat; during moult, the birds have to call on their fat reserves and then the DDT was liberated and had killed them. The strong winds would have the effect of blowing them onto the coasts and into oil-slicks that were about at the time.

At the second meeting at NERC it was announced that in the bodies then examined, polychlorinated biphenyls (PCBs) had been found in considerable amounts. This is tantamount, although many may not agree, to an indictment of the PCBs as the cause. It is possible that the birds had been feeding on fish living in contaminated water and had been so weakened by these substances that they were unable to survive the combination of stress through moulting and the lack of food through the strong winds. Without the initial weakening, under normal circumstances the birds would have been quite able to overcome both these factors, as in fact they do year after year and have done so for thousands of years. It has been said that a combination of factors was operating to the disadvantage of the birds at the time of the disaster, but the only thing that can be said definitely about the combination was that it brought the bodies in vast numbers up on the shores of western Scotland and northern England and eventually on most of the coasts bordering the Irish Sea. It will probably never be known whether the birds could have got over the poisoning had these other factors not been operating, but the significant factor is that *before* the winds began to blow, the birds had been seen feeding and in seemingly good condition, and had been observed dying soon afterwards.

Throughout the investigations and long after, there was, to my mind, a dangerous attitude; the attitude that it is a shame but these things will happen, have happened before and will happen again, that this was a natural occurrence, which may or may not have been aggravated by PCBs. Very few considered that it was the other way round, that it was the PCBs which were the prime factor and the rest aggravated the situation. What could have been the consequences for us? Probably none; at least, nothing has come to light. It might be expected that we should have been able to find fairly high concentrations of these chemicals in the fish that these birds were feeding on. Samples have been taken and so far the amounts of PCBs in them are not sufficient to account for the build-up in the birds, but these samples may not have been representative and furthermore there is just a possibility that the birds took in the PCBs directly from the water while drinking. We must not ignore the fact that, although this is a local incident, PCBs have been found in varying concentrations in food fish from the Baltic and the North Sea. The lesson to be learnt from this kind of incident is that the experts must be got together quickly to examine all the information that is available at the time and that they must go all out to find the cause and to announce it publicly, no matter how unsavoury the result may be. There is far too much at risk to cover up for fear of starting a witch-hunt. Action has already been taken by one manufacturer of PCBs which will go some way to ameliorate the sitation.

How the polychlorinated biphenyls enter the water is important, but of more immediate importance is the fact that they are there, and this is just as applicable to other pollutants. However, if we are to prevent pollution, we must know where they come from. There are fairly well documented mechanisms which show how industrial and agricultural wastes get into the marine environment. The first is, as we have seen, by way of drainage down the rivers. The second is by direct run-off from the land, the third by deliberate dumping and the fourth through the atmosphere.

In 1969 the coastal waters of Monterey Bay in California were contaminated by DDT running off from the land in heavy rain. This bay, although it is wide open to the Pacific Ocean, has rather sluggish water so that pollutants finding their way into it remain there for some time. Into this bay runs the Salinas River which drains an area of 4,000 square miles of the Salinas valley. For most of the year there is no flow through the river and its mouth is blocked by a sand bar, but during the rainy season from December to April the river flows and breaks through the sand bar into the bay. During exceptionally heavy rains it may overflow its banks and flood thousands of acres of farmland. When this happens the river turns into a muddy torrent and during February 1969 it discharged more than 100,000 cubic feet per second of muddy water that stained the sea for as much as three miles from the coast. The Salinas valley is a rich agricultural area and to keep its productivity high 125,000 lb. of DDT have been used on it each year over the past ten years, so that during the heavy discharges early in 1969 the silt was carrying heavy concentrations of chlorinated hydrocarbons. During the late spring and summer of that year, dead sea-birds were found along the shore of the bay in numbers far exceeding anything seen before. On one stretch of beach less than a mile long there were over 400 dead birds. Of these, 37 per cent had been oiled, 14 per cent shot and 49 per cent had died of causes unknown. Scientists examined the livers of this third group for chlorinated hydrocarbons and found that they contained concentrations of DDT residues as high as, or higher than, had been recorded anywhere. In one ring-billed gull 805 parts per million of DDE were found. In addition to the birds, many sea-lions were found dead on the beaches, and a pair of very sick Californian sea-lions which died later had from 4·0 to 89·0 parts per million DDE in their livers. And as Professor E. C. Haderlie, of the Department of Oceanography of the Naval Postgraduate School, Monterey, said in his report: 'We cannot at present be certain that it was the high concentration of DDT residues that killed these animals for we haven't data on the tolerance levels or lethal doses for any of these species. But circumstantial evidence seems to point to the pesticides as a cause of death.' This caution on the part of scientists is understandable, but the question is being asked more and more, do we have to wait until we have the proof down to the last detail before we take action? Do 10,000 Liverpudlians or Glaswegians have to die before we admit that a chemical is dangerous?

The third way in which the sea is polluted is by the deliberate dumping of materials, and this has caused a great deal of controversy, especially when the Americans announced, again in 1969, that they were going to dump 27,000 tons of lethal nerve and mustard gases into the sea off the eastern United States coast. We will discuss this particular incident later, but every day ships leave the ports of the industrial nations with the sole purpose of dumping poisonous wastes into the open sea. In the case of Europe, if this dumping goes on much longer the European fisheries will be ruined. The Norwegian Government is seriously concerned about the alarming rate at which this practice is increasing. Most of these wastes are the by-products of, once again, the plastics industry, and they include 1–2 dichloropropane, the ubiquitous chlorinated hydrocarbons and compounds of bromine. Some of these are known to be highly poisonous to most forms of marine life. Experiments have shown that as little as 10 milligrammes/litre of dichloropropane can kill cod and other full-grown fish, and it only takes a fraction of this concentration to destroy the larval stages and plankton. One case of dumping which was investigated revealed that a Swedish firm was dumping chlorinated hydrocarbons from its vinyl chloride plant at Stenungsund into north Norwegian waters, which are one of the main spawning grounds of the Arctic cod. Fortunately the enlightened Swedish Government stopped this practice and the factory is now storing its poisons at Stenungsund until it finds a better way of getting rid of them. Unfortunately there are few governments which act so promptly.

Ships regularly leave British ports ostensibly to dump factory wastes in deep water, but there is no control on the movement of these vessels; recently a seaman who had retired from one of the vessels employed in this 'trade' admitted that in bad weather they dumped not far out to sea, in this case the Irish Sea, and 'cooked' their logs accordingly. This was reported to the Ministry of Agriculture, Fisheries and Food, but as far as I know it was not substantiated and therefore no action could be taken. But the allegation concerned dumping of materials, including PCBs, in the north Irish Sea, not long before the 'Great Sea-Bird Wreck' discussed above. Coincidence or not?

Norwegian fishermen frequently report the finding of metal drums and canisters in their nets, and practically every time these contain chlorinated hydrocarbons and 1–2 dichloropropane. In the summer of 1969 the Norwegian research vessel *Johan Hjort* on a cruise in the North Sea from Jutland towards the Scottish coast sailed through a large area of dead plankton and dead and dying fish. The belt, which was several yards wide, was over seventy nautical miles long. The scientists on board

believed that they must have been sailing in the wake of a
vessel that was discharging its wastes into the sea. Specimens
of the dead fish contained chlorinated hydrocarbons. The
scientists took a serious view of this because whether the material
came from factories on the continent or in Great Britain, it had
been dumped in one of the main North Sea spawning areas for
mackerel. It is known that very large areas of the North Sea
have been used for dumping and one scientist said recently
that the sea-bed is 'virtually covered' with drums of chemical
waste nearly all of them containing chlorinated hydrocarbons.
Mr Grim Berge, a Norwegian scientist, said after pointing out
that many of the drums in the North Sea were leaking, 'This
means that if these drums have been dumped all over the North
Sea, their contents will slowly but surely be released to poison
the waters around them'. If this is so, and we have no reason
to doubt the extent of the dumping claimed because many of
the captains of the ships involved have provided the information,
then the North Sea fisheries will very soon be a thing of the past,
and probably those of the Irish Sea will disappear at the same
time.

Meanwhile sea-birds are dying in their tens of thousands all
over the world. Unaccountable deaths of sea-birds and seals
have been reported from the Aleutian Islands, South America,
South Africa and all over Europe. These reports are the norm,
not the unusual nowadays. Many species will be wiped out long
before the end of this century. We cannot take these sea-bird
deaths lightly, for they show that something is radically wrong
somewhere. They feed on fish that our commercial fish eat and
it is more than likely that the materials, toxic or otherwise, that
are being absorbed in these will ultimately cause a breakdown
either in the fish themselves or in their predators, including
Man. Apart from the effects, if any, on human beings there is
a danger that they might interfere with the genetical mechanisms and, therefore, the future of a species.

This may have happened already. The plague of starfish now
munching its way through the coral reefs of the Pacific and

Indian Oceans may have been caused by pollution. Whatever the cause, the plague is going to cost us most, if not all, of the hard corals of the South Seas, the Indian Ocean and possibly even the Red Sea. The only safe place, so far, for these corals is the West Indies. The 'Crown of Thorns' starfish, or *Acanthaster planci*, is normally a fairly rare animal; it lives off the polyps, that is the living parts of a coral. The hard corals are those that build the coral reefs like the Great Barrier Reef of Australia and the coral atolls. The first signs that something abnormal was afoot happened way back in 1962 when large concentrations of the starfish were seen off Green Island near Cairns in Queensland, Australia. Four years later nearly 6,000 individuals were counted in a 100-minute survey on one small section of this part of the Great Barrier Reef. Not far away on the Reef near Fitzroy Island the starfish began to swarm in August of the same year when eighty were seen. By March 1967 there were well over 4,000, and by October it was all over. They had almost gone for the simple reason that they had completely devoured the coral. It was dead and beginning to break up. The rate of progress along the Great Barrier Reef has been fantastic —one starfish can denude over 100 yards of coral a night—but *Acanthaster planci* has given up its normal eating habits, it no longer chomps its way through the late evening and early morning, it now eats all day!

The 'Crown of Thorns' plague has spread rapidly throughout the Pacific. It has destroyed over half the Great Barrier Reef. In two and a half years it has eaten away 90 per cent of the coral of the island of Guam. It has radiated out and reached the islands and reefs of Micronesia, Hawaii, Fiji, Borneo, the Solomon Islands, Midway, New Caledonia and Samoa. It has passed through the Malay Straits to the renowned reefs of Ceylon and is believed to be reaching plague proportions in the Red Sea and the South Indian Ocean round the Seychelles.

Some scientists believe that the hard corals which form the reefs and atolls of the tropical seas are bound to become extinct. This will certainly be a blow to the tourist industry and to the economy of the islands, many of which depend for their income almost entirely on tourism. Where the reefs begin to break up, as they do when the polyps die, the sea will begin to erode the islands and coasts away. Some of the beautiful coral islands, including many that are inhabited, will disappear.

What caused the sudden explosion in the population of *Acanthaster planci* is not known. Some marine biologists believe that it is a normal event, not unlike the occasional population explosions of the lemmings, and will die down eventually, and that the corals will recolonize the area when the starfish has passed. There are plenty of coral larvae floating about in the plankton and there is sufficient food for them, but they cannot recolonize because the sea breaks the old coral up, preventing the larvae from settling. Among a great many scientists there is the uneasy feeling that somewhere along the road Man is responsible, and there is some justification for this uneasiness.

The fact is that the 'Crown of Thorns' seems to have been more abundant near towns and where there is known pollution of the water. How the pollution has affected the starfish is anybody's guess. The abundance of food provided by sewage may have enhanced the survival rate of its larvae, or enabled them to avoid being eaten by their predators (maybe they do not taste so good in polluted water!). Anyway, far too many of them are surviving and reaching the adult stage. Another factor that might have some bearing is that shell collectors have practically cleared the islands of the adult starfish's principal predator, the triton. But it is unlikely that this is more than a secondary factor.

There is another possibility, which I haven't seen suggested before, and that is that a mutant form of the starfish may have appeared with characteristics that enable it to survive normal predation and which changed its habits. This moving about during the day, for instance, might prevent the triton attacking it; maybe the triton hunted and killed *Acanthaster* when it was resting during the day. If it *is* a mutant, it might have occurred naturally—this does happen; or it might have been caused by man-made chemicals in the sea-water—lead, mercury, PCBs, DDT, they are all there; or it might have been spawned by atomic radiation. The radioactive fall-out from the nuclear explosions which the Americans, the British and latterly the French have carried out in the Pacific has injected enormous amounts of radiation into the ocean. These could have caused any number of mutations in the marine fauna and flora. Luckily most mutations are bad and put the owner at a disadvantage in the environment, but some provide advantages giving the bearer a lead over the normal members of the species so that its progeny will eventually displace them.

The plankton in the waters near Bikini Atoll is radioactive and the coconut crabs living on the island are still accumulating enough of the radio-isotope strontium 90 in their shells to make them unsafe to eat, while the giant clams living on the sea-bed near the islands are so radioactive that the geiger counters were useless when scientists tried to record the extent of their contamination.

Whatever has triggered off the plague of starfish must be traced and, if humanly possible, dealt with. Unfortunately not everyone agrees on the seriousness of the threat, but the facts are, at the time of writing this book, that three million starfish are working their way over the outer fringe of the Loadstone Reef only thirty miles from Townsville. They have eaten their way up the side of the coral from the deeper water and are beginning to spill over into the shallow waters of the lagoon. The twelve-mile-long Arlington Reef has been reduced to rubble in three months. How far does disaster have to go before the authorities are convinced that it is time to act? On the American islands scientists helped by the United States Government are working on the problem round the clock. But this is a Pacific-Indian Ocean problem. It needs an international effort to deal with it before it is too late and the corals and much of the beauty of the marine world have been obliterated.

International co-operation regarding the regime of the sea is an ideal, but one which is much easier to talk about than to put into practice. It is difficult enough to get agreement on regulations concerned with the international waters of the oceans, but virtually impossible if they involve the so jealously guarded territorial waters. And, as we have seen in the case of the United Kingdom, it is in these inshore waters that pollution can be particularly heavy. If the pollution stayed in the local waters it would be bad enough for those who live near by, but it would at least present no threat to others. Unfortunately water doesn't stay put and there is no guarantee that pollution will be only local. One form of pollution which is causing much concern is the atomic waste from nuclear power stations which is discharged either directly into the sea or in river estuaries.

There are eight nuclear power stations round the British coasts, each of which is discharging its share of radioactive pollution into the coastal waters. Whether this is having any effect on the marine environment is questionable at this stage, but what of the long term? Presumably these power stations are only the first of the many that will be needed to provide the power necessary to keep the country going, so we can expect that the discharges will increase and will go on pouring into the sea indefinitely. Already the United Kingdom has six more nuclear power stations under construction and five more on the drawing board. Multiply this waste by the discharges pouring into the coastal waters of other countries and it adds up to a large injection of radioactive material. And in common with all materials that get into the waters of the continental shelves, and in particular the shallow seas round the coasts, these materials will enter and be accumulated in the food chains. Radioactive salts are known to cause cancer and to affect the genetic composition of a species and to produce mutations, and in time we may have plagues of new forms of marine animals, a repeat of the 'Crown of Thorns' in another guise. Alternatively, it is quite possible that we shall experience in time the complete destruction of a species which may well be an important commercial fish or a vital link in the food chain. The outstanding thing about our knowledge of the effect of radiation on living organisms is our ignorance. We do know for certain that it attacks the very basis of life, the DNA molecule, and one questions the wisdom of allowing any nuclear wastes to run freely into the waters around our coasts.

More particularly we should question the decision that allowed the planning for six nuclear power stations on the shores of the Bristol Channel, and another which is under consideration at Stourport farther up the River Severn. The argument in favour of the Bristol Channel as a centre for concentrated nuclear effort is that it is within a convenient distance of the industrialized areas of South Wales and the west Midlands. There is also a very large tidal flow (in fact the tidal range is one of the greatest in the world) and an abundant supply of cooling water. But for all that the Bristol Channel is confined and, as in all estuaries, the water will wash to and fro, allowing time for the radiation to enter the life-cycles. Admittedly these waters are, or should be, of low radioactivity. The high-level radioactive wastes are 'stored', which means that they are put into concrete silos or into stainless steel tanks surrounded by concrete and dumped into some deep hole in the continental shelf or into the oceans.

But it is not only the nuclear power stations that discharge radioactive wastes into the sea. The Atomic Energy Authority's factory at Windscale for processing fuel for the nuclear power stations disposes of its waste through a two-mile-long pipe into the north-east Irish Sea. In May 1970 concern was expressed by the Ennerdale Rural District Council about the UKAEA's application to increase the amount of α-emitting waste discharged from 450 to 2,000 curies a quarter (the curie is the standard unit used to measure radiation). The Council, supported by the Cumberland County Council, naturally expressed its anxiety about the possible harmful effects of the accumulation of the wastes. The waste from the plant consists of chemicals used in processing fuel, which become radioactive, but only plant-washings of the short-life radioactive nuclides are disposed of in this way and their rapid decay in the sea, coupled with the dilution, are considered sufficient to make the effluent harmless. However, the spread of the wastes has been tracked by the detection of tiny traces of radioactive substances in some of the marine algae, or seaweeds, to the south of Windscale and northwards into the Solway Firth. Ennerdale RDC showed particular wisdom in seeking help from a non-government body in assessing what this increase might entail. The Council approached the British Society for Social Responsibility in Science and asked them if they would look into it. As it turned out, the BSSRS Committee of five physicists recommended that the council accept the request of the UKAEA and allow the extra discharge, but the decision applied only to that one specific request, and any other increases would need further study. As the *New Scientist* said: 'Although there was no confrontation, it is comforting that alternatives to silent acquiescence do exist.'

The discharge of the 2,000 curies per quarter is still very much below the internationally accepted limits, but even if they were not, who would know about it if the British Government wanted to keep it to themselves? On the face of it the British arrangements are encouraging. Before any radioactive wastes can be disposed of into the sea, authorization from the appropriate ministries has to be obtained and this can be given only when those ministries are satisfied that the discharges are well within the safe limits. Areas round the British coasts where power stations and other nuclear establishments discharge radioactive wastes are monitored for damage to the marine life. In a written reply to a question in the House of Commons in February 1970, Mr James Hoy, Joint Parliamentary Secretary, Ministry of Agriculture, Fisheries and Food, said that his ministry 'is satisfied that there have been no measurable effects of radioactive effluents from nuclear power stations on marine life. In addition to the regular control and monitoring of discharges into the sea, the ministry's scientists undertake extensive investigations into the level and effect of radioactivity in the marine environment.' Nevertheless, there is nothing toprevent the United Kingdom or any other country from signing a convention regulating disposal of wastes, while at the same time doing just what they like in their own waters if it is convenient. There is no way of monitoring, on an international scale and in a disinterested manner, what is discharged into the sea, or of ensuring that agreed regulations are adhered to.

Nuclear power stations are not the only ones to be sited on the coasts. To meet the massive demands of modern society for energy, economics dictates that many conventional power stations must also be built on the coasts because of the cheap supply of cooling water. When this water has been used in the power plant, nuclear or conventional, it is returned to the sea several degrees warmer than when it was taken out. It is polluted by heat and this thermal pollution can influence the ecology of the sea, rather like the situation described for fresh-water bodies.

The advent of desalination plants will bring its own crop of biological problems. These water distilleries will pump water out of the sea and use the 'cheap' electricity generated by the nuclear power stations to produce fresh water. This has got to be done, particularly in countries like those in the eastern Mediterranean and the Near East where there is an acute shortage of water already, and eventually, if they don't put their fresh-water resources in order, in the industrialized countries, and again in particular in the United Kingdom and the United States.

Taking into consideration the enormous quantity of water in the sea, it seems unlikely that the pumping of a few million gallons of water a day can have anything but a very minor effect, but it should be remembered that the ocean mass is a solution of many salts, not just sodium chloride. It is this fine balance which is important to the living organisms that have evolved in this 'soup', and if we start altering it, even ever so slightly and locally, we might be inviting disaster by depriving them of the complete environment to which they are adapted. The trouble is that with all these projects the economic considerations come first, the engineering second and the ecological not at all. We can no longer afford the luxuries of technology at the risk of injuring the environment, even in the sea. The seas and sea bed can provide a lot of the resources that are becoming scarce on the land, provided they are exploited with care. The shallow seas are, as we are constantly reminded, the most important. They are already the sites of the first palaeotechnological attempts at extracting minerals and other resources. One of these resources is oil.

Oil pollution of the sea is the worst maritime pollution problem we have to solve and yet technically it should be the easiest. It is a problem that affects the United Kingdom particularly. In order to maintain our standard of living, these islands have to import every year 100 million tons of oil which is brought by tankers thousands of miles across the sea. Many more tankers, bound for refineries in other parts of north-west Europe, habitually sail close to our shores, and so it is inevitable that oil will somehow or other find its way into our coastal waters and those of Europe's continental shelf as a whole. The problem of course is not just a European one, it occurs wherever oil is transported.

Oil can get into the sea in a number of ways: through accidents, deliberately—when ships clean out their tanks or bilges—through human error or mechanical damage, and through the leaks which occur at oil terminals or from oil rigs. The classic case of an accident at sea, and one that really focused attention on to the oil pollution problem, was the wreck of the *Torrey Canyon* in 1967. When this vessel, with its load of 117,000 tons of Kuwait crude oil, hit the Seven Stones Reef a few miles west of Land's End six of her tanks were ripped open and immediately the oil poured into the sea. Soon a slick of oil approximately eight miles long stretched southwards from the stricken tanker. The story of how *Torrey Canyon* was bombed does not need re-telling. But the significance of the *Torrey Canyon* and its destruction lies in the fact that we were not prepared to cope with such a disaster, with the result that over a hundred miles of Cornish coastline were covered with a thick mass of crude oil. The damage to amenity was obvious; rather less so was the damage to marine life, which was underplayed.

The blobs of black oil that foul the beaches and the layers of 'chocolate mousse' that coat our shorelines ruin clothes and shoes, stick to the skin and in general spoil holidays, and may eventually ruin a multi-million pound tourist industry. This is bad enough, but the real threat from oil is to marine life. It had been assumed until quite recently that the oil itself had little or no effect on the biology of the sea and seashore, and after *Torrey Canyon* most of the damage to the shore life was due to the chemicals used to clear away the oil. Today we know that the oil itself can cause widespread damage to marine organisms directly and indirectly by reducing the solar radiation penetrating the surface of the sea and the gas exchanges between the atmosphere and the water.

But the sea-birds have taken the brunt of the assault by oil pollution. The total number of sea-birds killed by oil pollution since it began can never be calculated. In the case of the guillemots, the birds that suffered most from the PCBs, it is fair to say that they will probably join the list of extinct species by 1980. These birds suffer most from oil because of their habit of diving rather than flying away from obstruction. They spend most of their time at sea, sitting on the surface or diving and swimming under water in pursuit of their prey, mostly small fish. This makes it difficult for them to avoid oil-slicks, and when they come into contact with one the guillemots, unlike gulls which tend to fly away, dive, and as many will surface in the oil-slick as outside it. There seems to be no avoiding action on their part other than by this random diving and swimming. It has been suggested by some writers that the birds are attracted to the oil-slicks because they may resemble food—often when there are shoals of small fish just under the surface the water has a peculiar texture which is not unlike that caused by oil, and thus high-flying birds and those on cliffs might be attracted to the slicks. But, as mentioned above, gulls, and these include most high-flying marine species, avoid oil-slicks. The cliff-dwelling species, such as guillemots, would be reasonably safe if the oil were to be found in the waters off their breeding cliffs at that time of year when the birds were in residence. But guillemots, the other auks and many diving ducks are killed throughout the year by oil, and a slick cannot be identified as such by a bird sitting on the surface of the water, at least not until it bumps into it. Dr Kai Curry-Lindahl of Sweden suggested that long-tailed ducks, which suffer badly through oil pollution, may land on oil slicks because of their calming effect on the water.

Whatever the reason, the fact remains that literally millions of sea-birds are being killed throughout the world by oil pollution. The Royal Society for the Protection of Birds reported an estimated death toll of 22,000 sea-birds round the coast of Britain in the first three months of 1970. In July 1970 Kenneth Williamson, the RSPB's representative in Anglesey, estimated that 100,000 sea-birds had died in the Irish Sea alone already that year. Extrapolating from these and other figures, a realistic estimate of the total mortality of sea-birds in the north-east Atlantic—that is, around Europe—in 1970 would be 380,000; the total number of sea-birds which will be killed by oil pollution during 1971 throughout the world is estimated to be 1,200,000 if the conditions prevailing in 1970 are maintained.

The guillemots have a continuous record of mortality due to oil, or which can at least be presumed to be due to oil, from the turn of this century onwards, with a peak around about 1917 and another during the 1939–45 war and a sharp annual increase ever since. This matches closely the rise in oil-burning and oil-carrying ships. The war peaks were due to sinkings of tankers and other shipping. Calculations based on these trends, and allowing for the total populations estimated to be living in 1960, which seems to have been the last year in which stability of populations was observed, show that the guillemots of the north Atlantic will be extinct by 1980, the other auks— razorbills and puffins—very soon after, and that shags and cor-

7 'The Norwegian explorer, Thor Heyerdahl, reported that he had encountered pollution for almost the full length of his journey across the Atlantic in his papyrus boat *Ra II*'—lumps of asphalt-like material taken from the Atlantic.

8 Mystery scars and ulcerations on fish caught in the north-east Irish Sea, scene of the 1969 sea-bird disaster.

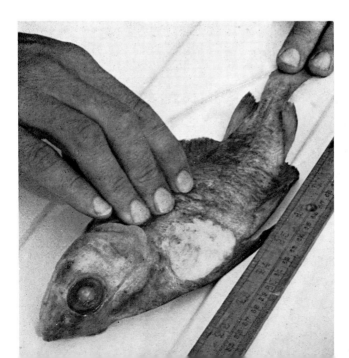

9 The effect of air pollution on vegetation. The bean plant on the left, grown in pollution-free conditions, is healthy, while the one on the right bears the scars of atmospheric pollution.

10 Fluorides from the chimneys of a phosphate factory were responsible for the crippling of this cow.

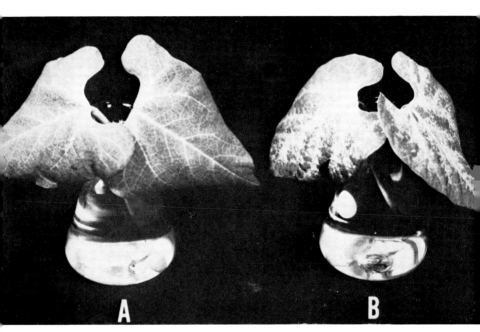

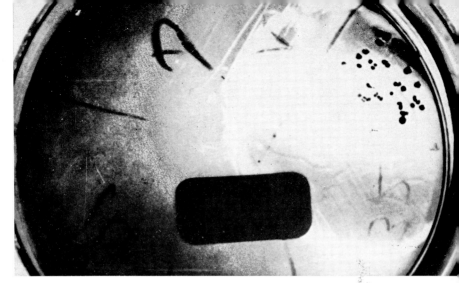

11   One organism at least, *Streptococcus mitis* (black dots—top right) is known to have got to the Moon on a television camera despite stringent sterilization and quarantine procedures.

12   As far as we know, *Homo sapiens* and *Streptococcus mitis* are the only organisms to have existed in Space, but this should not lull us into negligence over quarantine procedures.

13 The ecological consequences of the Aswan Dam are already being felt on the fertile delta of the Nile (dark area) and the nearby Mediterranean.

14 The exploration of Space has impressed upon us the smallness of our home and emphasized the problems we face, for the limits of our planet have been sharply defined.

morants will have joined them by 1984. A recent count of shags in north-west Scotland showed losses of 85 per cent—on the Island of Canna in 1970 the population was reduced from 1,000 to 150 pairs.

The slaughter of birds, although deplorable, may not be of great importance from the view of the marine resources but, as we have said before, it is an indicator of the deteriorating condition of our coastal seas. If the cause of guillemot deaths in the early part of the century had been recognized, something might have been done about oil pollution earlier.

Oil itself kills, but the chemicals used to get rid of it not only kill but may also affect the marine organisms by interfering with their reproduction. The thousands of tons of detergent used to treat the crude oil that spilt out of the *Torrey Canyon* destroyed and maimed many of our shore-dwelling animals. The detergents do not stay in the surface layers of the sea but disperse throughout the mass of water. Two years after the *Torrey Canyon* disaster and the subsequent cleaning of the beaches with detergents, the dog-whelk population along the south Cornwall coast had not recovered. A study of these whelks around Porthleven, where 100,000 gallons of detergent were used to disperse the oil, showed that although there were only a few of them, they were the only species to survive. But their numbers recovered quickly after their food animals, barnacles and marine worms, had re-populated the area. A population explosion of dog-whelks occurred, followed by a general decline because their predators—crabs and small fish—did not recover and balance the population structure; the whelks, uninhibited by predation, soon devoured all their food. Another effect of the detergent was to produce deformities in the surviving dog-whelks which, if they are inherited, may in the long run put them or their descendants at a disadvantage in the environment.

The great difficulty is that we don't know what these chemicals can do. The word 'detergent' is used to cover a large variety of products, from shampoos to oil emulsifiers, and so it is almost impossible to know what their actions will be either individually or in various combinations in the sea. All we can say is that most of them, and certainly those used after the *Torrey Canyon* disaster, are poisonous to marine animals at very low concentrations in the water. Many animals perished when exposed to 10 parts per million for only a couple of hours. The main cause of this toxicity was known to be the aromatic solvent in the detergent which was necessary to deal with the viscous oil. The dispersant was applied to the shore neat, and consequently played havoc with the marine life. Similar materials entering estuaries and shallow waters could destroy fishing grounds and wipe out mussel and oyster colonies. It does seem, however, that the shore-dwelling fauna are more resistant to the constituents of the detergents than are those that live below the low-water mark. Some of the solvent emulsifiers, for instance, are poisonous to adult mussels, oysters and cockles at between 10 and 100 ppm within twenty-four hours, while shrimps and related crustaceans are killed by lower concentrations in a third of the time and their larvae by 1 to 20 ppm in three hours. These chemicals must be considered to be every bit as dangerous as oil pollution itself.

Unless something is done rather quickly oil and detergents are likely to be the primary agents in the extinction of many organisms, including the sea-birds, in the shallow seas. We should think twice before allowing extensions of oil terminals and oil extracting operations in the sea. The disaster of the Santa Barbara oil rig leak is a salutary lesson for the exploiters. Here in 1969 the drills punctured a natural fault in the ocean floor and released a quarter of a million gallons of crude oil into the surrounding sea and onto the shore. A year after the explosion the oil was still coming out, destroying marine life and ruining the Santa Barbara beaches. Such a disaster could happen anywhere oil is being extracted, and the present exploration on the Great Barrier Reef, which, as we have seen, is already under assault from a plague of starfish, should be viewed with caution. In the North Sea, one of the richest fishing grounds in the world, we have seen a quickening of its death through dumping and the run-off from the rivers of the surrounding countries. It now has to contend with oil and gas rigs. If a repeat of the Santa Barbara disaster were to happen in this area, then there would be little doubt about the consequences.

The destruction of the shallow seas through pollution is no longer fantasy, it is fact. Already the Baltic, Black, Caspian, North and Irish Seas are polluted to varying degrees. The combination of the pollutants from the Finnish and Swedish wood-pulping industries, agricultural run-off, the sewage and other wastes from Leningrad, Copenhagen, Malmö and Stockholm has created a very serious situation in the Baltic. There has been a noticeable drop in the oxygen content of the deeper waters, and if this continues the entire water mass will probably become devoid of life, a situation which seems to be developing in the Black Sea. Phosphate concentrations in the Baltic are now three times higher than they were in 1955 and estimates give a figure of around 400,000 tons of phosphates in the deeper water, which is being added to at a rate of 16,000 tons derived from sewage and industrial wastes every year. The DDT concentration in the Baltic seals is ten times higher than that found in North Sea seals. Scientists are now closely watching to check whether this has any effect on the reproductive ability of the animals and whether there is an abnormal increase in mortality. Scientists have also been trying to find out how the oxygen content of the deep waters has been reduced. It is known that the lowering of the oxygen concentration is related to the production of hydrogen sulphide which has been found in those waters. It seems that the phosphates from the sewage, detergents and agricultural and industrial wastes stimulate the growth of marine plants which absorb the oxygen from the water, and after their growth cycles are over their remains sink to the bottom and decompose. This generates the hydrogen sulphide which is trapped below the halocline (in bodies of water such as the Baltic there is at certain depths an acute change in salinity, which prevents the mixing of the waters above and below it—in the Baltic this is at a depth of 180 feet) where it concentrates and destroys all life.

If we continue to pollute the seas as we are doing at present, then all the shallow seas will suffer a similar fate; and yet we expect them to provide more and more of our food. At present about 10 per cent of the world's protein comes from fish. The world's fishermen crop between sixty and seventy million tons of fish a year. By the end of the decade their target will be nearer 200 million tons—but will pollution allow them to reach this? Relentlessly it is pushing its way farther and farther out from the land. Beyond the edges of the continental shelves lies the deep water of the oceans, out of sight, out of mind.

CHAPTER 9

The deep, the wide and the forgotten

During a cruise in the western Atlantic in 1970 the American research vessel *John Elliott Pittsburg* dredged the five-mile-deep Puerto Rico trench. In one haul the scientists discovered a small fish only 6 inches long. It turned out to be the deepest living backboned animal ever recorded from such a depth. This little denizen of the deep, known to the scientists as *Barrogigas*, was accompanied in the dredge by empty bottles and cans, pieces of aluminium and disused flashlight batteries. That the deep oceans had become rubbish dumps had been evident for some time, but no-one really imagined that they were as squalid as this!

The Norwegian explorer Thor Heyerdahl reported that he had encountered pollution for almost the full length of his journey across the Atlantic in his papyrus boat *Ra II* in 1970. Most of the pollution he saw consisted of solidified 'asphalt-like' lumps ranging in size from pieces as small as peas to some as big as potatoes. Sometimes the pollution was so bad that they couldn't use the water to wash their teeth, and this in mid-Atlantic! This muck was more than likely the rubbish left from oil-burning ships and tankers, which had dumped it into the sea. From this account it would appear that the whole of the north Atlantic was covered in oily wastes.

But before we jump to conclusions of this sort we must examine the facts a little more closely. To the crew of *Ra II*, the world consisted of what they could see from the boat—a very small area of sea bounded by their horizon. They were being carried along on the North Equatorial current, which sweeps across the Atlantic from North Africa to the Caribbean. Were they seeing the same sheet of water and its complement of pollutants, as they drifted on the ocean current? If so, it would give them the impression that they were sailing through mile after mile of polluted water. And it must be remembered that they were cruising in one of the major oceanic currents, which would be carrying any pollutants it picked up from the Straits of Gibraltar, Portugal, north-west Africa and the accumulation of rubbish carried down from the tangle of sea lanes passing into the Mediterranean and to northern Europe. As it happens Heyerdahl kept a log of *Ra II*'s speed and collected samples of the pollutants and had them analysed—the results showed that they were passing through different pollutants as they cruised along. The waters of the ocean keep together in discrete masses, and the argument could have been put to Thor Heyerdahl if he had not kept records that he had been seeing the same pollution day after day. But the argument turned the other way round also holds good for those who claim that materials dumped into the sea would be immediately dispersed; they wouldn't, they are more likely to hold together for quite a long time before dispersing.

One good example of this is the case of the copper which killed thousands of fish off the Dutch coast in 1965. Strictly speaking, the particular incident belongs to the shallow seas, but it is used here to illustrate the point, for it might equally have occurred hundreds of miles from land. Nobody knew in this case what was killing the fish until a chemist found copper sulphate crystals on the beach and subsequent analysis of the sea-water showed that the copper content was five hundred times greater than normal. Whoever had dumped the copper offshore had probably expected it to be dispersed, instead of which a mass of copper-rich sea-water had been formed. This discrete mass of water moved northwards and was still killing fish near the entrance to the Wadden Sea and threatened its rich mussel beds. Luckily north winds blew up and pushed the water away from the coast, and in time it dispersed.

I will discuss the deep-water dumping of obsolete nerve gases in a later chapter, but it is pertinent to mention here that the graveyard for a large quantity of American nerve gases is the region where the Florida current gives birth to the Gulf Stream.

If such material were to escape from its concrete coffins and did not hydrolyse, as the authorities assure us it would, but were to remain lethal, then it would find its way into and be carried by the surface currents up the Carolina coast, to join the main Gulf Stream, eventually washing up on the coasts of Europe and contaminating some of the richest commercial fishing grounds in the world. So much then for the advocates of ocean dumping!

The ocean is a mass of currents and counter-currents. Many of those below the surface are uncharted, many remain unknown. The waters are never still, but are pushed relentlessly on by a combination of wind and the Earth's rotation. In the northern and southern hemispheres the Trade Winds blowing from the north-east and south-east respectively are responsible for the great North and South Equatorial currents flowing westward across both the Atlantic and Pacific Oceans. On reaching the coasts of the continents these massive currents are deflected northward in the Northern Hemisphere and southward in the Southern Hemisphere, forming a great clockwise and anti-clockwise movement. Between these Equatorial currents are found the counter-currents that flow towards the east. Many of the warm equatorial currents connect with the cold-water currents flowing in from the polar regions. The Gulf Stream consists of the deflected North Equatorial current upon which *Ra II* sailed. A look at the map will show that any persistent pollutants drained into the Atlantic from the American east coast rivers will lap up on the coasts of Canada, Newfoundland, Greenland, Iceland, the British Isles, Norway, France, Spain and Portugal. The stragglers, if they miss the North Africa coast, the Azores, Madeira and the Cape Verdes, will go swinging round and come up into *Ra II*'s current to the Caribbean, or back where they started. The Gulf Stream is responsible for those bright beans and coconuts which from time to time are washed up on the shores of the west coast of Ireland, the Scillies, Cornwall and Devon. These tropical beans, which are sold as souvenirs, originate from the north coast of South America or the islands of the Caribbean. Nowadays it is just as likely to be a piece of garbage as a cocoa bean which makes the journey. Conversely, the muck draining from industrial Europe can find its way to all stations from Guyana through the British Virgin Islands to New York. In the centre of this vast whirl of water lies the strange Sargasso Sea with its unique flora and fauna, the 'sink' of the Atlantic Ocean. Here, if anywhere, persistent organic compounds will find their way to be absorbed in the life cycle of the Sargassum weed. What an opportunity to look for oceanic pollution! Here we might have a built-in indicator. Even if tests on animals were negative it would provide a base-line to work from in the future.

In the Pacific, pollutants coming off the western American seaboard enter the California current, then cross the tropical Pacific into its North Equatorial current (the longest current in the world) to be deflected by the land mass of south-east Asia, up past the Japanese islands in the Kuroshio current, the equivalent to our Gulf Stream, and back across to America in the North Pacific current. It is easy to see that the pollution in both hemispheres not only slowly disperses, but keeps on going round and round in the ocean region in which it was dumped.

Very slowly the waters do mix and contaminated waters eventually enter the adjacent currents. Gradually they seep into all the oceans and in the end go circling round the Antarctic in the West Wind Drift. If we had only the surface currents to deal with, it would make some sense to dump dangerous materials in the deep trenches of the oceans. But the oceans are moving throughout their depths, so they do not constitute a 'safe' deposit bank for the waste from our military arsenals and nuclear power stations, if the deep currents are to bring them back to the surface in a few months or years. If Lord Delacourt-Smith meant what he said in a parliamentary answer in May 1970: 'radioactive salts are being stored in the sea, not dumped', we should ask the question, for whom are they being stored? Our generation, the next, or the one beyond that? Some of these radioactive nuclides remain dangerous for thousands of years.

The deep currents in the oceans move relentlessly on, some slow, some fast, and where they meet head on they force the water from the bottom to the surface in great upwellings. It is in these areas that the oceans are rich in salts brought up from the seabed. These provide the nutrients for the plankton, and because of this it is in these regions that rich oceanic life is found. The radioactive and other materials that we are now dumping in the deep oceans will, if they leak out, eventually find their way into these fertile areas in precisely the same way as the natural salts. They would enter the food chains and destroy them for ever. If this should happen—and we have already seen that the other rich area of the Earth's oceans, the shallow seas, are under threat—we could lose the productivity of the entire oceanic mass. An example of one of these upwellings is the Antarctic convergence, where the cold water from the Antarctic ocean slips under the warm water coming from the tropics. Here, because of the upwelling of the salts, one of the richest areas of the world's oceans is found. The richness of the plankton supports a staggering variety of life, an abundance of fish, sea-birds, seals and the great whales, a living treasure that Man has voyaged for years from Europe to glean.

It is not easy to understand why the ocean, which after all is a fluid medium like the atmosphere, should have been thought less mobile than the latter. They are interdependent. The heat of the ocean drives the atmosphere, and in turn the movement of the air drives the currents. One of the most vital areas of the world is where air and water meet, that narrow band at the bottom of the atmosphere and the top of the ocean. For it is here that oxygen produced by the phytoplankton, which some scientists have calculated contributes some 70 per cent of the Earth's oxygen supply, enters the atmosphere. The importance that nothing should interfere with this interchange is obvious. This is why observations like those made by Thor Heyerdahl and his crew have to be taken seriously. The lumps of tar themselves are not as important as that fraction of the oil that spreads out thinly over the surface of the sea, which, if it was extensive enough, could reduce and eventually stop altogether the photosynthesis of the phytoplankton, depriving it of carbon dioxide and to some extent of sunlight and preventing the escape of oxygen into the atmosphere. The same thing of course applies for any pollutant that spreads over the sea's surface in the same way. Pollutants in the atmosphere also cut down the solar radiation and thus diminish the efficiency of the phytoplankton and therefore the productivity of the oceanic food chains.

Scientists working on board the oceanographic research vessel *Atlantis II* on a voyage through the Mediterranean and eastern North Atlantic in 1970 found thousands of black and brownish lumps of tarry material on every square kilometre of the sea's surface. The *Atlantis II* was cruising from Rhodes in the eastern Mediterranean to the Azores in the North Atlantic and the towing nets became fouled with the lumps of tar, which were frequently covered with barnacles and other crustaceans. From the measurements of these barnacles the scientists were able to estimate the age of the tar lumps, which turned out to be at least two months. It is possible that the bacteria which were also found in the tar would eventually break it down, but how long that would take nobody knows. But it is not only the spoiling of the sea's surface that worries the scientists. More important is the fact that the fish living on the barnacles are also taking in the oil and so this is entering into the marine food chain. The disturbing observation was that many of the oil nodules contained the lighter, poisonous fractions of petroleum locked inside them. Several of the fish that were caught during the cruise were found to have large amounts of the tar in their stomachs. And these fish form the food for many of the commercial fish, such as tuna, used by Man. The worst area for the pollution, according to their observations, was to the south of Italy where the amount of tar reached some 500 litres per square kilometre. The presence of the light petroleum fractions seems to indicate that these lumps originated from fuel oil and not from ships' bilges, as was thought at one time. Some scientists believe that these lumps, which form a chronic pollution problem, may significantly affect the whole marine ecosystem, especially in an enclosed sea like the Mediterranean.

All these observations so far have been in areas where one would expect to find high concentrations of pollution, and of oil in particular. They are in areas of busy shipping lanes and adjacent to continental masses. Already we have seen the destruction that oil pollution has wrought in the shallow seas, but now we are beginning to see the threat it holds for that 71 per cent of the Earth's surface, the oceans. Tankers of a quarter of a million tons are becoming commonplace, half-a-million-ton tankers are on their way and already one-million-tonners are being planned. The marriage of economics and technology is posing one of the most difficult problems that face mankind, for as long as petroleum is used to provide energy it will have to be shipped or piped from country to country across or under the seas. Accidents will happen and leaks will occur. It has been estimated that something like a million tons of oil are lost in this way every year. With the arrival of the supertankers this will increase out of all proportion. It takes only a small amount of oil to cover a wide area a few molecules deep.

The oceans are immense: they cover most of the Earth's surface with an average depth of about 5,000 feet. They contain 360 million cubic miles of water and it is hard to grasp that the puny amount of man-made pollutants can have any effect at all. Each cubic mile of sea-water without Man's additions contains 165 million tons of dissolved material, the mineral resources of the sea. How then can our activities possibly affect this mass of water? In the first place we can destroy the prime producers of energy, the phytoplankton which form the basis of the oceanic food chains. If we do this we should lose not only the protein that the fish provide for the human race but also the necessary oxygen. In the deep sea the productive regions, as we have already seen, are limited to a few specific areas where currents carry the minerals to the surface. Off the coast of South America the famous Peruvian or Humboldt current supports a vast and rich fishery upon which Peru depends for its livelihood. It is imperative that pollutants, whether they be radioactive salts, war gases or just plain rubbish, must not get into these areas to be mixed with the nutrient materials. If they do (and DDT and its derivatives are already universally found) they will be absorbed into the food chain and will be concentrated as they penetrate up the chain.

If the pollutants were to become sufficiently concentrated to alter the density of the water, it would have such an effect on the life of the sea that either a whole new regime would be set up, probably to our disadvantage, or marine life would be destroyed altogether. If the density increased, the plankton would be concentrated at the surface and over-crowding would occur, and although this would eventually sort itself out, damage to the plankton might be severe through the intensity of the solar radiation being too strong and causing a collapse in their photosynthetic processes. Alternatively, if the water density were reduced (this is very unlikely) the phytoplankton would sink out of reach of the energy altogether and bring about total demise of life in the sea.

Another danger, and one that concerns all life in the sea, but particularly the small organisms found in the plankton, is the lowering of surface tension in the cell membranes. The action of detergents, domestic or otherwise, is based on their property of lowering surface tension, and the fact that millions of tons of these substances are constantly flowing down our rivers into the sea must cause some concern, because as they are being absorbed into the food chains they could very well reach a concentration at which all the physiological processes dependent on the transmission of fluids through cell membranes would be destroyed. In the case of the phytoplankton, detergents in the surface waters could very easily rupture their cell walls and kill them. A small patch of detergent at sufficient concentration could destroy literally billions of planktonic organisms.

The chances of the complete disappearance of life in the sea seem at this date to be so remote as to be hardly worth considering, but then who would have thought in 1945 that DDT would be found in penguins in the Antarctic Ocean?

It is popular to speculate upon the mineral wealth of the oceans. Much time and energy are being devoted to the exploration of these resources, and it can be expected that this activity will increase as the land resources are exhausted. Perhaps too much emphasis has already been put on the value of the food resources of the sea; on their own they will never be able to supply sufficient protein for the human population, for even the vast productivity of the ocean is limited. And even with more efficient fishery technology, including fish farming, the percentage will become smaller in the future as the human numbers grow. Nevertheless it will be an important portion and will have to be maintained and improved. To achieve even this, the present rate of pollution will have to be halted and any further plans for marine exploitation, whether for minerals, power or just plain water, will have to be carefully thought out and implemented with the minimum interference with the living systems. It is vital in our own interests for us to see that this is done. None of the decisions should be left entirely in the hands of the oceanologist, and they should never be taken with only economic or political considerations in mind. There must be co-operation at all stages in any future exploration or development of the oceanic resources between the technologist and the ecologist.

Except to those few of us who have their business on the waters, the oceans seem remote, but, of every hundred breaths we take, seventy come from the ocean.

CHAPTER 10

The winds of change

In a small way Man began polluting the Earth's gas envelope when he first started lighting fires; even today, among the worst polluted atmospheres are those found in the huts of the 'primitive' peoples of the world. Atmospheric pollution is the oldest form of pollution and was early recognized as such, but more in terms of its nuisance value than as a danger to health. But Man got used to it, and quietly it got worse until today it rates as a major threat to all life on this planet.

Statements such as 'A new Ice Age would threaten civilization if the cloud cover of the Earth were to increase by only 6 per cent' may seem to smack of exaggeration and sensationalism, but they do make us wonder whether such a thing could happen and also, momentarily at least, make us think about the atmosphere. We are more conscious of it when a gale howls around us or when our surroundings are shrouded in fog. We all become acutely aware of it when we run for a train or a bus and drop panting into our seats. We are panic-stricken when we are deprived of air for only a few moments. It might take weeks to starve to death, days to die of thirst, but it takes only seconds to suffocate! Yet despite this, the fouling of the atmosphere is still the most difficult problem we have to solve. Most of the difficulty derives from apathy and meanness, but before we go into these let us look at the atmosphere.

The Earth's atmosphere is a well-balanced mixture of gases which we call air, extending some 600 miles from the Earth's surface into space and thinning out with increasing height. By volume 78 per cent of air is taken up by inert nitrogen, 21 per cent by oxygen and the remaining 1 per cent by a number of other gases, some of which, as we shall see, can have very important effects. Nitrogen acts as a fire-blanket; without it almost all living tissue would burst spontaneously into flames. But by far the most important of all these gases is oxygen, without which life, except for a few specialized forms of bacteria, cannot exist. Deprive the human brain for two or three minutes of its oxygen supply and it will be irreparably damaged and death will soon follow. This simple element is produced by green plants during photosynthesis, and in fact before the green plants evolved the Earth had a reducing atmosphere which consisted of gases such as methane, ammonia and carbon dioxide. It was the primitive plants using carbon dioxide and the energy of the Sun which, by liberating oxygen produced during photosynthesis, gave us our oxidizing atmosphere and made animal life possible. A great deal depends on this oxygen, for not only are the normal 'ticking-over' life processes dependent on it but its availability enabled nature to produce a great variety of animal species.

The chances of Man interfering with the oxygen profit and loss on a world scale seemed remote a decade ago. Today many scientists are having second thoughts, and although no such threat hovers over our heads in the immediate future, the danger cannot be discounted altogether. Some 3,500 million human beings (and the number increasing all the time) each breathing 7 lb. of oxygen every day, the burning (for oxygen is needed to support combustion) of the colossal quantities of fuel needed to provide them with heat and to drive the machinery producing their material requirements, the exposure of oxidizable materials, such as iron, all contribute to the debit side. In addition to this continually increasing loss, we are also reducing the profit side by our destruction of the world's forests and other photosynthetic areas of the land and, as we have seen, of the sea. Whether the productivity of the remaining green plants will continue to provide a surplus is not known, for like a good many of the large-scale environmental factors, we do not know what the profit and loss are in real terms. The difficulty too is that we should be unable to detect any adverse trend on a global scale until it had gone some way in its development.

The destruction of the Earth's green mantle is bound to bring about climatic changes, for transpiration, the evaporation of water from the leaves of vegetation, is an important factor in maintaining the Earth's humidity. Changes have already been observed where vast areas of forest have been cut down and this destruction of forests and jungle will go on—it must, to make room for the increasing numbers of humans and their food crops, which are usually less efficient both in moisture conservation and oxygen production than the wild inhabitants they have displaced.

Not only do we depend upon the plants for our oxygen but we also rely on them to absorb a large amount of the carbon dioxide from the air. All living things, including the plants themselves, respire, that is, they take in oxygen, which is used to convert food material into new tissue and energy, and expel the waste product from the process, carbon dioxide. When we fill our lungs with air, our blood quickly absorbs the 21 per cent of oxygen, and in return carbon dioxide and water vapour are expelled when we breathe out. This exchange of gases takes place deep in the lung tissue in little sac-like processes called alveoli. In these, the capillaries bring the deoxygenated blood from the tissues, and because the haemoglobin (the pigment that gives the blood its characteristic colour) has a greater affinity for oxygen than the carbon dioxide it is carrying, it makes a swap through the very thin semi-permeable membranes of the alveoli. When the water vapour and carbon dioxide are expelled they are freely available in the atmosphere, and during daylight hours the carbon dioxide is readily taken up by the green plants. The photosynthetic process needs far more carbon dioxide than the plants produce during their own respiration. At night, with no light available to keep the photosynthesis going, the carbon dioxide is not absorbed and the plant's own respiration adds to it, but in nothing like the quantities to justify the old belief that plants should be taken out of a bedroom at night! However, at least this superstition did show that people were aware of the plants' role and importance in the environment, an awareness which seems to be evaporating today. There is, therefore, a direct exchange of gases between plants and animals, another piece of the environmental jigsaw, a dialogue between plant and Man. It is hardly surprising to find that in areas where carbon dioxide is allowed to build up, such as in busy city streets and in some industrial premises, people are not so healthy. To get sufficient oxygen out of the air, and in such places the oxygen percentage will be lowered, their hearts have to work harder to push the blood through the alveoli at a faster rate. The consequences of this increased heart rate may be reflected in the growing incidence of heart failures.

With more and more people and less and less green cover, the carbon dioxide must be building up in the atmosphere. This is very much news today, for it could give rise to severe climatic changes. While it is fairly easy to visualize that high concentrations of this gas locally in the environment may give rise to ill-health, it is difficult to imagine its effects on a global scale. To understand the effects that it could have, we must know the mechanisms whereby the Earth and its atmosphere maintain the ideal conditions under which life has evolved and continues to exist.

The planet, and therefore all life, obtains its energy from our star, the Sun. The energy given off by the Sun is dissipated into space and some of it arrives at the outer shell of the Earth's atmosphere. This blanket of gases absorbs those wavelengths which would be damaging to living organisms, and only a small part of the solar radiation gets through to the surface of the Earth, where it is absorbed by rocks, soil and sea. Some of the energy is reflected back into space, but at longer wavelengths, and this mechanism keeps the Earth from overheating; in other words, it maintains a thermal balance. If there is a build-up of carbon dioxide in the atmosphere, these long waves cannot escape because they are absorbed by the carbon dioxide. This means that the Earth's temperature will rise and if it continues to do so by only a few degrees, it might become hot enough to melt the ice-sheets of the polar regions. This is often known as the 'greenhouse' effect, so called because the glass in a greenhouse behaves in a similar way to the carbon dioxide, that is, it lets in the visible wavelengths of the light, absorbing some on the way, but does not allow the longer wavelengths generated by the soil and the plants to escape, and so the greenhouse heats up.

In all natural systems there is a balancing effect; in this case the energy enters in one form and is re-radiated in another, thus maintaining the energy balance. Likewise, the gases in the atmosphere are in constant proportions according to temperature and pressure, and you cannot go on pumping more in without there being a compensating absorption somewhere else. So if we are pumping carbon dioxide into the atmosphere, we must be losing something which is making room for it. In the natural gas budget of the planet, the oxygen and carbon dioxide are in equilibrium. But with the advent of industrial processes, combined with the destruction of vegetation, an imbalance in this equilibrium is apparent. Interference with a natural order might have far-reaching but inter-connecting effects. There is no evidence at the moment to suggest that on a global scale the proportions of atmospheric gases are changing, but locally there *are* changes and some of these are causing concern.

As far as carbon dioxide is concerned, on the global scale we are safeguarded by the fact that so much of the Earth is covered with water, which has an affinity for the gas. In consequence much of the surplus carbon dioxide is being absorbed by the oceans and presumably will continue to be so until the oceans themselves become saturated with the gas. The present fears of sweeping changes in the climate may seem far-fetched, therefore, but when the limit of the oceans' capacity to absorb carbon dioxide is reached, then these fears may well be justified. We do not know how long this would take, we do not know even whether it would happen or whether some other compensating mechanism would come into play. For example there is some evidence that the increase in carbon dioxide has stimulated the plants to increase their activity. Thus the vegetation and the oceans are protecting us from having an excess of carbon dioxide. In spite of the local build-up of carbon dioxide in the thirty years since 1940 the global average temperature has not risen as some have forecast; on the contrary, it has dropped by about 0·3 deg. C. Whether or not this change in average temperature is due to natural alterations or whether it is itself an expression of another form of pollution produced by Man is not known. After all, fluctuations in temperature have been known before, Ice Ages have come and gone, and even in the time of recorded history the ice-sheets of Europe, Greenland and Iceland have advanced and retreated. At the height of the Viking expansion, Greenland and Iceland were experiencing what must have been for those countries a warm period, for the Norsemen were able to colonize both countries. Later the communities in Greenland were destroyed principally by climatic deterioration, while today, once again, Greenland has on its west coast flourishing mixed Eskimo-European communities. But in the past ten years the trend has been once again towards a build-up of ice, so much so that the Icelanders have experienced an increase of ice on their northern coast and harbours and have had difficulty with their fisheries and harvest. It is hard for us to differentiate between natural fluctuations in temperature and climate and those that might possibly be caused by pollution, and only long-term studies will provide the necessary information.

The only thing we can say is that the rate of change seems to be rather rapid. By and large, natural processes of change are slow—an alteration may take hundreds or thousands of years because of the sheer size and volume of the atmosphere, but Man's influence, although on this scale seemingly insignificant, can have a profound effect on the rate of change, as we saw in the case of the ageing of the Great Lakes.

This latest cooling of the Earth's atmosphere could have been caused through pollution. One theory put out to explain the phenomenon is the presence of particulates—particles of solid or liquid substances ranging in size from those visible as soot and smoke to those so small that an electron microscope is needed to see them. They can remain in the air for long periods and can travel considerable distances. They are mostly derived from the burning of fossil fuels, but also include natural dust which may be made up of materials such as sand, salt and volcanic ash, and man-made dust from industrial processes and agricultural chemicals.

The effect of these particles suspended in the air is to lower its transparency to the Sun's energy, and they can increase cloud formations by providing the nuclei around which moisture can condense. This would have the effect of cooling the surface of the Earth and would to some extent cancel out the heating due to the increased carbon dioxide. It may go even further, and it has been estimated that a decrease in atmospheric transparency of as little as 3 to 4 per cent would produce a drop in temperature of about 0·4 deg. C. An addition of only 1 per cent in the global cloud cover would lower the temperature by 0·8 deg. C. This is nearly three times the drop measured in the last twenty years. Normally about 31 per cent of the Earth's surface is covered in low clouds. If this went up to 37 per cent, the temperature would fall steeply by about 4 deg. C., which would bring us very near to the conditions that would send us plunging into another Ice Age. So the statement at the beginning of the chapter is not perhaps as alarmist as it seems, for although in practical terms such a situation may be improbable it is not impossible. However, there is no evidence to suggest that the average cloud cover of the Earth is increasing, and there is certainly no proof that pollution is the cause of this recent drop in temperature, but it could be having some effect, and if the present rate of increase in atmospheric pollution in the form of particulates is maintained or accelerated, it is bound to have some effect. But it is more likely that natural dust, mainly from volcanic eruptions, or a combination of natural dust and man-made particulates is responsible. An erupting volcano can throw out half a million tons of ash every hour during its activity and much of this finds its way into the upper atmosphere, where it may float about for years. However, it must be admitted that there is no evidence to substantiate this thesis: volcanic eruptions have not increased in frequency during the past seventy years or so. But if it turns out that man-made particulates plus the volcanic dust are responsible as suggested here, then it would be yet another example of Man acting as a catalyst.

Regardless of whether man-made particulates are having any significant effects on the atmosphere as a whole, they are certainly the cause of a great deal of suffering to mankind. The most numerous particulates in the atmosphere to date have been the products of the incomplete combustion of fossil fuels, and smoke and soot are the worst offenders. Smoke is probably the oldest atmospheric pollutant and has always been a nuisance. As long ago as 1273 the smoke problems were so bad in London that an attempt was made to ban the burning of coal, and in 1661 John Evelyn made a special attack on the same evil in his pamphlet *Fumigorium*. With the arrival of the Industrial Revolution in the nineteenth century, the rot really set in, and the overriding drive for industrial development at any cost swamped all the protests at the pall of smoke which hung over the cities and towns, in particular London. Great Britain has by no means been the only sufferer: all industrialized countries have experienced the same problem. The peak of smoke emissions was reached between 1930 and 1960.

The smoke problem is aggravated still further when there is sufficient water vapour in the air for the soot particles to combine with the water vapour to produce 'smog'. Smog was a common occurrence in the 1940s and early 1950s, and London was particularly prone to it. The worst smog ever recorded hit London in 1952 and was responsible for the deaths of 4,000 people, and the ill-health of thousands of others. This at last triggered off enough public reaction to force the Government of the time to take action, and this led to the Clean Air Act of 1956. Since then a number of improvements have been made to extend the power of local authorities to introduce smoke control regulations, and in some enlightened cities and towns—London and Sheffield in particular—smog has become a thing of the past. In Greater London smoke control is estimated to cost 15p per person per year, a small price to pay for a cleaner and healthier environment. One would have thought that such an example would have been followed by all local authorities, and certainly a reduction in the total smoke emissions from 2·25 million tons in 1956 to 0·87 million tons in 1967 is a measure of the success of smoke control, but there are still twenty-three local authorities which have done nothing about applying it. Most of these are to be found in the north of England, an area which suffers twice as much smoke as the south. There is a hardened attitude here against smokeless zones, difficult to explain except in terms of bonds of loyalty to coal-burning.

In the little Yorkshire town of Rawmarsh, where an estimated sixty-two tons of grit, sulphur and smoke are deposited on the town and in the lungs of the inhabitants every year, the resistance to smoke control is very strong, as was shown in an article in *The Times* in March 1970, in which one of Rawmarsh's civic leaders was quoted as saying:

I just don't believe in smokeless zones. There is nothing like a good coal fire. I just do not believe that domestic coal causes damage to health. I am convinced that all the diseases they attribute to coal are contracted at employment, especially in the heavy industries round these parts. Rawmarsh will give careful consideration to smokeless zones when they are considered necessary. Until then we shall stay as we are.

One wonders when smokeless zones are considered necessary! But Rawmarsh is by no means the only place in the United Kingdom that has this attitude to clean air zones. Aspull near Wigan in Lancashire is another. In this case the excuse is cost. In the same article cited above, a councillor, another retired colliery worker, is quoted as saying that Aspull could not afford smokeless zones, and went on to say: 'We are on high ground here, you see, and the prevailing winds are from the direction of Southport on the coast. They blow plenty of fresh air across Aspull!' It would appear that the councillor is not in the least worried about where his pollution goes after the fresh air from Southport has blown it from his backyard! This attitude is an example of some people's ingrained resistance to change and is particularly noticeable in the mining communities, where the traditions of coal-burning die hard. There is a fear that new-fangled methods and fuels might damage the coal industry, but it might be argued that this is a left-over from the bad old days which preceded the Second World War, a sociological problem which should, we hope, die out with that generation.

No doubt officialdom is very satisfied with the overall results of its legislation, but how many times do we see it negating its own efforts, the classic 'one hand not knowing what the other is doing'. The Ministry of Technology, for example, removed its subsidy from the Northern Gas Board's smokeless-fuel producing plant on Tyneside. Without the subsidy, the Board found that it could not operate the plant economically and was forced to shut it down. The result was a shortage of smokeless fuel on Tyneside, and Newcastle-upon-Tyne, Sunderland and Hebburn had to apply for a suspension of their smoke control orders. This was a body blow for enlightened and community-minded people who had fought hard to combat the social evil of smoke pollution. *The Times*, 25th March 1970, quoted Alderman Dr Malcolm Thompson, Chairman of the Newcastle Health Committee, as saying: 'We have met a crisis in our smoke-control programme caused by the withdrawal of the subsidy. It is a disastrous state of affairs and very disappointing to those of us who have worked for cleaner air. . . . Someone has blundered. A decision has been made without regard for the consequences of environment and health.'

The shortage of smokeless fuel is not confined to Tyneside. The long-term planning of the requirements brought about by smoke control has not matched the need. Lord Robens, then Chairman of the Coal Board, forecast the shortage of clean fuels for the winter of 1970–1 and he asked the Government not to introduce any further smoke control zones for two years. The shortage of smokeless fuels, while not perhaps bringing us back to square one, will certainly put a brake on the drive for cleaner air, at least as far as smoke is concerned. The domestic chimney is still responsible for about 85 per cent of the pollution of air by smoke in the United Kingdom, and even if plenty of smokeless fuel were available, there are many people, particularly the old, who just cannot afford it and must either burn coal or freeze.

The Clean Air Act is concerned mainly with reducing and controlling the emission of smoke, grit and dust, and places limitations on the density of smoke that can be emitted from the chimneys of buildings, from railway engines, etc., and includes provision for the setting up of smokeless zones. That it has done, and is doing, much to control the emissions of visible pollutants is very evident, but this is only one step in the right direction. The Act does not make provision for the invisible pollutants which are pumped into the atmosphere, for although air may look clean, this does not mean to say that it *is* clean.

The burning of fossil fuels, coal and oil, produces the waste gases sulphur dioxide, carbon monoxide, nitrogen oxides and carbon dioxide, and these are still among the major threats to human life. It is not certain that they produce diseases in themselves (although there seems to be a growing mass of evidence to suggest that they may do so), but without doubt they aggravate and sometimes trigger off diseases already existing in the population. Bronchitis is one of those complaints aggravated by pollution from both particulates and gases, especially sulphur dioxide, so much so that in 1965 between 35 and 40 million working days were lost in Britain through it, and it still costs this country some £28 million every year. In the smog of 1952 it was this disease intensified by the smog that killed 4,000 Londoners, and it is still recorded as the cause of death in Britain more frequently than anywhere else in the world. No wonder it is called the 'British Disease'! As might be expected, the greatest proportion of bronchitic sufferers are to be found in the industrial areas of the Midlands, the north of England and London, and a similar correlation has been observed between areas with high pollution levels and the incidence of rheumatism and arthritis. All these pollutants might also have a connection with lung cancer. Soot and other solid particles are taken into the lungs, where they block or irritate the delicate lung tissue. This lowers the efficiency of the organ and puts great stress on the respiratory system and the heart. In addition to this, the tarry matter in suspension in the air can irritate the skin, the nasal passages and the eyes. Atmospheric pollution also reduces the amount of sunlight getting through, which accounts for the pallor of the urban dweller, to say nothing of the psychological effects which the lack of sunlight has on him. For the same reasons, it lowers the efficiency of photosynthesis in plants, and clogs up the stomata in the leaves and thus can kill them, particularly the evergreens. Deciduous vegetation is not so badly affected because this can grow new leaves every year, but in time these too will be affected.

Sulphur dioxide is the most widespread of the invisible pollutants: it is a heavy gas, very pungent and poisonous, and is produced chiefly by the burning of coal, coke and oil in refineries, chemical plants, power stations and melting and smelting operations. In Great Britain the quantity emitted each year has increased from 4 million tons in 1938 to 6 million tons in 1967. The proportion from coal has remained more or less static, and the increase is mainly due to the great expansion in the use of oil. In the free state sulphur dioxide combines with the oxygen and water in the atmosphere to form sulphuric acid. This is very corrosive and eats into the stonework of buildings and decorative and functional metal work, cars, aeroplanes, bridges, in fact anything that is exposed. It also attacks paper and cloth. The cost of this damage in Britain is about £1 million a day and in the U.S.A. it is $5 million a day, but this does not take account of its damage to health. It has already been mentioned that combined with particulates, sulphur dioxide can seriously aggravate bronchitic conditions, but exposure to even low levels of it can cause shallow rapid breathing and an increase in the pulse rate. As sulphuric acid, it penetrates deep into the lungs and damages the tissues. A very disturbing possibility is that sulphur dioxide may be a contributory cause of congenital defects. Dr Robert Shapiro, Associate Professor of Chemistry at New York University, claimed that as sodium bisulphate it reacts sharply with RNA and DNA, two body acids playing a dominant role in heredity. These findings have been supported by Dr Frank Mukai at the same university, who has obtained mutations in bacteria by using sodium bisulphate.

Plant life provides a good indicator of the amount of sulphur dioxide in an area. Lichens and bryophytes are absent in urban and industrial areas because of their extreme sensitivity to some air pollutants, particularly sulphur dioxide, and with some lowly forms of plant life it is even possible to construct a scale showing the levels of sulphur dioxide present. It has been suggested that when the annual average concentration of sulphur dioxide exceeds about 50 microgrammes per cubic metre then the productivity of higher plants, conifer trees to lettuces, can be affected. Sulphur dioxide from a large copper smelting plant set up after the American Civil War in Copper Basin, Tennessee, was blamed for the destruction of 30,000 acres of timberland, and most of this area is still barren today.

How can we get rid of it? As most of it comes from fuels, the logical answer would be either to burn fuels which do not contain any, such as natural gas, or to treat the fuels prior to burning or the waste gases after burning. Natural gas is certainly being used increasingly, but the cost of converting equipment to it is high. So too is the cost of removing sulphur from coal and oil. Coal and coke can be 'washed' and oil is refined, but not the heavy fuel oil which is most often used. Without any controls, 1 tonne of coal burned in a power station can emit up to 9 kilogrammes of sulphur oxides, 8–10 kilogrammes of nitrogen oxides and 10–20 kilogrammes of particulate matter, while natural gas would give off only very small amounts of particulates and sulphur oxides and only up to half the amount of nitrogen oxides. The Americans have designed a system, 'Cat-Ox', for extracting sulphur dioxide from flue gases before they go up the chimney. The gas is converted to sulphuric acid which can be used for the manufacture of phosphate fertilizers and ammonium sulphate, but even with this to offset the cost, the system is very expensive. Furthermore it is designed for installation in new plant, and it would add 20 per cent to the cost of building a power station.

Economics have dictated that so far the only thing to do is to push the sulphur dioxide up into the atmosphere out of chimneys built as tall as possible. These often have to have special boilers to heat the fumes so that they can shoot out sufficiently fast for them to be carried up and away. But what goes up must come down, and this applies to all the pollutants shot out of chimneys in this way.

Atmospheric pollution is by no means just a local problem, it is very much an international concern and no one can be completely isolated from its effects. Pollutants from brickworks in Bedfordshire have been suspected of doing damage in Stockholm, Sweden, and DDT has been found in the fat of penguins, seals and fish in the seas surrounding the Antarctic continent, carried by the atmospheric currents thousands of miles away from the nearest areas where it is used. The steadily increasing acidity in rainfall is probably caused by widespread air pollution.

Radioactive fall-out is an obvious danger, and each nuclear explosion increases the risk. Radioactive fall-out from early explosions is still present in the atmosphere, particularly the upper layers, and is gradually finding its way into fresh water and on to the land—and into our lungs. If the radioactive substances in the air contain alpha emitters—biologically damaging radiation—and these enter the lungs or the gut through eating and drinking, then the alpha emitters will expend all their energy inside the body and attack the blood, lung tissue, stomach, intestines and bone marrow, causing various radiation diseases and cancers, including leukaemia—a condition of the blood due to over-production of white blood cells. Such is the power of invisible air pollution, all the more terrifying as you may not be aware that you are being attacked.

In the United States, particulates and sulphur oxides contribute only 27 per cent of the total of 200 million tons of pollutants pumped into the atmosphere every year. Of the remainder no less than 47 per cent, or nearly 100 million tons, is the colourless, odourless and lethal gas carbon monoxide, produced primarily by the incomplete combustion of the fuel of vehicles, with the ordinary private motor-car the worst offender. If the total level of carbon monoxide is maintained at 1 per cent for only half an hour it can kill. On average, car exhaust gases contain 3·5 per cent of carbon monoxide, and this can go up to 7 per cent when the engine is idling, as when standing in traffic jams or warming up in a garage. This is why so many people die through carbon monoxide poisoning when stranded in their cars during blizzards. They keep the windows shut and the engine running to keep warm, and under these conditions the carbon monoxide can soon build up to the lethal concentration. The levels of concentration of this gas in the air close to traffic in busy streets can be as much as 25 parts per million. This is six times the accepted maximum allowable concentration for an eight-hour exposure, and 10 parts per million over this period is sufficient to retard mental reactions. It is no wonder that policemen in Tokyo have to have oxygen administered to them after a spell of traffic duty! The new Sony Building in Tokyo has an oxygen vending machine available from which for a small charge you can have a revivifying breath of 'fresh' air. Carbon monoxide combines with the haemoglobin in the blood thus depriving the body, and especially the brain, of oxygen. We have already seen that it takes only a little while for death to follow a stoppage of oxygen to the brain.

Vehicle exhausts also contain lead, an extremely poisonous substance. This is added to petrol, usually as a mixture of tetraethyl and tetramethyl lead, in quantities up to approximately 1 millilitre per litre, and its purpose is to act as an antiknock agent, to prevent the unpleasant effect of explosive burning ahead of the flame front in the combustion chamber. In recent years concern has been growing at the unexpectedly high blood-lead level shown by people not in contact with the substance industrially, and it has been suggested that the lead emitted in vehicle exhaust fumes could be responsible. Lead affects the nervous system and particularly the brain, and thus there might well be some correlation between it and the sluggishness observed in some drivers and pedestrians who have been in areas of high concentrations of motor-car fumes. But to differentiate between the various gases and other pollutants, and to state which is responsible for what, is almost impossible.

Under certain conditions different gases can react and form secondary pollutants which can be equally, if not more, dangerous. Under the action of sunlight, nitrogen oxides, of which a large proportion derive from vehicles, and hydrocarbons, of which more than half come from the incomplete combustion of fuel in petrol engines, react to form what are called 'photochemical oxidants'. This family includes ozone, nitrogen dioxide, peroxyacyl nitrates, aldehydes and acrolein, and when present in the air they can irritate the eyes and lungs and cause damage to plants. Combined with solid and liquid particles they produce photochemical 'smog', and it is this type of smog for which Los Angeles is notorious. The problem is intensified if a 'temperature inversion' condition prevails.

Under normal conditions the air is colder the higher up it is, but sometimes a reversal of this takes place and we get what is called a 'temperature inversion'. In this the upper air is warm and traps the lower layers. This 'upsidedownness' of the atmosphere frequently occurs when the sky is clear, the air very dry, and there is hardly a breath of wind. This set of conditions encourages very rapid radiation of the heat from the ground and the lower levels of the atmosphere. The cool air at the bottom is very dense and so it stays near the ground while the upper air loses its heat only slowly. Consequently the pollutants which are pumped into the lower levels of the atmosphere from chimneys and exhausts are trapped, and when this occurs the visibility goes down and people have the discomfort of having to breathe in relatively undiluted pollutants. In Los Angeles temperature inversions are fairly common, and as that city is surrounded on three sides by high mountain ranges trapping the air even more, the choking fumes persist, sometimes for days, even months. Although Los Angeles may be the most chronic sufferer, many other large cities experience these condistion. New York and Tokyo had a particularly bad time during the summer of 1970. The London smog of 1952 was the result of a temperature inversion trapping smoke and particulates, but this is not a common occurrence—the prevailing winds are generally too strong and the hills surrounding the London basin are relatively low.

The pollution of the Earth's atmosphere is now a fact of life and there is nowhere on the surface of the planet where we can escape it. Locally it might be more intense, but inevitably the atmospheric currents carry the pollutants to every quarter of the globe. We live at the bottom of the atmosphere just as the deep-sea fish live at the bottom of the sea, and like them we are subject to the forces that govern our fluid medium. Whatever else we do, we cannot interfere, except to our cost, with this envelope of gas which sustains us and protects us from the storms of the cosmos.

CHAPTER 11
The space walkers

Above the tenuous protective layers of the upper atmosphere lies the unknown. After a journey of a quarter of a million miles and several million years of evolution, Man has penetrated space and has stood at last on another planet. On 20th July 1969 two men stepped onto the Moon. As one of them, astronaut Neil Armstrong, placed his foot on the lunar surface he said: 'That's one small step for a man, one giant leap for Mankind.' It was a spectacular jump in more ways than one!

From that moment, Man's exploration of space really began, although its great liberating message has yet to be understood by the majority of Mankind. For some, including the astronauts, it has impressed upon them the smallness of our home and emphasized the problems we face, for the limits of our planet have been sharply defined. Space exploration has also brought its own problems. Like the oceans, it seemed to a few that space might be a solution to, not part of, the problem of pollution. But even in his first ventures into this great uncharted province Man has taken his untidiness with him.

On 4th October 1957 the Russians launched their Sputnik 1 satellite into orbit. Already hundreds of obsolete artificial satellites, their ranks swollen with the abandoned carcases of the rocketry that put them and the lunar and planetary probes there, are circling the Earth. Some in time will re-enter the atmosphere and end their lives in an incandescent glow as man-made shooting stars. It is unlikely that space junk will represent any danger, although if it continues to grow it might present a hazard to astronauts. Nobody has given much thought as to whether there could be any consequences from its just being there, or whether, in fact, the ionization of returning probes and junk as they thunder into the atmosphere may over a long period of time have a cumulative effect on the protective layers of the Earth's upper atmosphere, those regions that protect us from the intense solar and cosmic radiation. In these regions much of the high-energy radiation is absorbed and prevented from reaching us.

In 1961 the Americans carried out what was called the 'West Ford' experiment. Strictly speaking, this belonged to the military, but as so much of the space work in the early days was done at their instigation, it is included here. The idea behind the experiment was that with a belt of copper needles they could create an artificial radio-reflecting layer. This, it was thought, once in orbit around the Earth at a suitable height where the resistance of the atmosphere is negligible, would act as a permanent radio reflector. If successful, radio communications throughout the world would be free of the blackouts caused by ionospheric disturbances when the Sun sends out its high-energy flares. What effects the needles might have on the layers of the ionosphere did not seem to concern the military. If the belt had become permanent the science of radio astronomy could have been throttled at its birth, but of more importance was the fact that no one knew for certain whether it would decrease the protective role.

The more responsible scientists were furious and they demanded that the experiment should be abandoned. Many nations joined in these protests, but the American politicians ignored the implications and the adverse world opinion and backed their military. On 21st October 1961 the West Ford experiment began and up went the needles. But to everybody's relief, outside the U.S.A. anyway, the experiment failed. Not to be outdone, another attempt was made, this time successfully, but the precious needles stayed up for only a short time and luckily no damage was done. So far no further attempts have been made and it is to be hoped that the American military has had enough of pins and needles!

Our planet is, as far as we know, the only polluted one, but if Mankind has anything to do with it there will be at least three more by the end of the century. We have already made a good start, for the Moon and Venus have their share of discarded metal, and Mars will soon be joining the club. To examine our polluting effect, we can imagine space to be divided up into a number of zones. The Earth itself forms the first of these and the earlier chapters have shown how a nearly hopeless situation has developed over the years. The next zone is the upper atmosphere or the fringe of space—this has already been violated by nuclear explosions, copper needles and space junk. Going out farther, we enter what is sometimes called near space, a few hundred miles out from the Earth, the graveyard of much technological rubbish. Farther out still, we enter the void of interplanetary space and at the end of the journey the planets themselves. One thing the scientists would like to know is whether there is, or ever was, life of any kind on the Moon, Mars or Venus. When the American astronauts first landed on the Moon, they saw no outward signs of life, but the samples of lunar rock and dust they brought back are being closely examined for any flicker of life or clues that might give us an idea of how life may have begun on Earth. But to most of us these questions, although interesting, are academic. What we want to know is whether the astronauts can bring anything back which could be useful, such as metals and other materials, or which could be harmful to us, our animals or our crops—questions which have inspired the science-fiction writers for decades.

To answer these questions it is essential that the astronauts and their equipment should be completely sterilized before they leave the Earth, and on their return, they, their equipment and any specimens they may bring back, must be quarantined. For Man's first trip to another planet, the National Aeronautics and Space Administration designed the Lunar Receiving Laboratory to meet these conditions.

Sterilization is a complicated business. The vehicle which is destined to land on another planet must not be exposed to the Earth's atmosphere before the launch. After sterilization it is put inside an outer casing which is ultimately jettisoned when the Earth's atmosphere has been left behind. Two methods used by the Americans to sterilize their space-craft are by bombardment with gamma rays, which do not affect the delicate electronic equipment inside, or by introducing ethylene oxide which is lethal to viruses and bacteria. Complex monitor systems check for any leakage before and after the flight. Nothing can get in or out without the trained observers knowing, in theory at least.

In practice, despite these stringent sterilization and quarantine procedures, one organism is known to have got through. Bacteria of the type *Streptococcus mitis* were found on the television camera of Surveyor 3, launched in April 1967 and brought back by the Apollo 12 astronauts in 1969. These organisms had managed to survive for two and a quarter years in the alien environment of the Moon, and if they had managed to escape the Earth's safety net, it is reasonable to presume that other space-craft have also taken their share of contaminants. *Streptococcus mitis* could have undergone mutations which might have turned it into a killer, producing diseases for which we have no natural or man-made protection. Elaborate precautions were taken with the Apollo astronauts when they returned from the Moon. The quarantine which they underwent was long and arduous, but unimaginatively linked to the twenty-one day period which most Earth organisms take to show themselves. Organisms from other planets might have shorter or much longer development periods.

However, *Streptococcus mitis* has given an indication of the kind of conditions under which life can survive in space. As long ago as 1908 Arrhenius, the Swedish Nobel prize winner, put forward the idea that life on Earth had been seeded from space, and he suggested that a primitive encapsulated organism may be able to float in space, tossed and driven by cosmic

storms, until it lands on a suitable planet to start the evolutionary process. This is one very good reason why it is so important
that we should not contaminate space or the planets. If such
'space-seeds' do exist they would certainly have landed on the
Moon and have remained dormant there, but if we contaminate
the Moon with Earth organisms we may lose the chance to find
out.

There is a further loophole in space-Earth contamination
quarantine. If cosmobiota exist in free space it can be sensibly
argued that any on the outside of a space-craft would be burnt
off on re-entry, but men have walked in space and returned to
Earth without any quarantine. This could even be made an
argument against space exploration altogether, but the possibility of contamination may be small and Man must take a risk
at some time—it is part of his story.

The only other planet to have been reached by a man-made
probe is Venus. The planet is a little over 24 million miles away
and nearer the Sun than we are; it is only slightly smaller than
the Earth and in many respects rather like our planet, but its
surface is completely hidden by a dense atmosphere, and until
Mariner 2, the first successful planetary probe, was sent there
by NASA in 1962 we knew very little about it. This probe by-
passed Venus at some 20,000 miles and sent back data showing
that the surface is very hot, and from this it has been deduced
that life, at least as we know it, is unlikely to exist there. Other
findings confirm this; for instance, ground level atmospheric
pressure on Venus is over ninety-seven times greater than that
of the Earth and there is very little free oxygen in the atmosphere, which seems to be made up mostly of carbon dioxide.
But in view of the conditions encountered and endured by
Streptococcus mitis it is not beyond the realms of possibility that
organisms could be living there and that we could contaminate
Venus with organisms from Earth.

The Russians have already successfully landed four vehicles
on Venus. It is to be hoped that they sterilized their probes,
but it will be impossible to know for certain now whether any

life or remains of life that may be found there did not have their origins on Earth, unless of course they are uniquely adapted to the Venusian conditions. Anything returning from Venus will have to undergo rigorous and prolonged quarantine checks. Mars, the 'red planet', is farther from the Sun than the Earth is, and thus it receives less radiation. Its surface temperature is around 15° C. by day and minus 73° C. at night. In spite of this, Mars has always been considered as an abode of life. The 'canals' discovered by Schiaparelli in the last century started off a lot of speculation and led to a new literature on the subject from H. G. Wells onwards. The passing within a few thousand miles of Mariner 4 in 1965 changed all this. This probe sent back pictures of the Martian surface, which turned out to be not unlike that of the Moon—pockmarked with craters. The atmosphere is thin and, like that of Venus, consists mostly of carbon dioxide. This thin atmosphere can give hardly any protection from the solar and cosmic radiation and thus life, if it does exist, is probably lowly. Even so, as with Venus, we must be prepared for any eventuality.

Many scientists think that we should have waited until sterilization techniques were further improved before launching vehicles to Mars and Venus. But Man was in too much of a hurry and, although precautions were taken, his impatience made him take a calculated risk.

The other planets—Jupiter, Saturn, Uranus and Neptune—seem a very long way off, but some time during the period 1976–1980, when these planets are in line (this happens only once in every 179 years) an unmanned space-craft will be sent to swing past them, a trip which will take some eight to eleven years. What Man does beyond that date will depend largely on what he wants to do, but he must think very carefully before committing himself or his equipment to further space exploration.

The men responsible for the space flights are fully aware of the dangers and can be expected to do everything possible to prevent contamination of the Moon and the planets and the infection of the Earth on return. We have the knowledge now that at least one Earth organism did survive the rigours of the launch, the space ride and the landing on the Moon, and lived there until it was picked up and given a comfortable habitat in the NASA laboratories. Our hurry to get into space has lulled our vigilance and maybe dulled our sense of responsibility and proportion. A mistake here could be more calamitous than any other of Man's activities. When the crash-landing of the second successful lunar module sent echoes round the Moon, one scientist immediately suggested that a nuclear device be detonated on the Moon so that he could learn about its structure. Fortunately good sense prevailed, but it is early days yet in lunar exploration!

The space junk in Earth orbit and the litter left behind by the astronauts, to say nothing of their sewage, is the beginning of pollution in space. It is sobering to reflect that the first world Man has visited beyond his own already has a garbage heap.

PART III

AN ENVIRONMENT IN MAN

CHAPTER 12

'Their own executioner'

We have seen how Man's activities have altered the environment and how those alterations in turn have affected his internal 'environment'. This is not surprising, for they have evolved together. After all, the Earth's atmosphere is contiguous with that inside Man's lungs. The oxygen of the air and that carried by his bloodstream is the same. The combustion that drives his cells depends on oxygen just as much as does the blast furnace in a steel plant. Flushed with the success of his 'conquest' of his habitat, Man has forgotten that he himself is an integral part of Nature; he does not realize that this conquest is undermining his own nature. Man, a product of the evolutionary battle for survival, by his mastery of so many of those forces that ensured that only the fittest of his kind would survive, has encouraged the weakening of his own species. The breakdown by modern medicine of those biological and environmental filters that allowed only the best to survive has produced a lowering of the physical and mental qualities of Man. The release of his reproductive stamina is not only draining the Earth of its resources but is diluting his genetical worth. Man is guilty of polluting not only his environment but also his genes.

Genetical pollution is a reality that few are willing to face. The saving of the lives of babies with defects caused by transmittable genetic faults can threaten millions of humans in the future. The ethics of modern medicine demand that the doctor saves the individual and, with the continuing advances in medicine and surgery, children with such defects are now able to grow up and reproduce themselves. And this is the rub and the dilemma. For in by-passing the natural filter we are making sure that those genetic faults are passed on and multiplied in future generations. The dilemma is, have we the right to save an individual, sometimes even before it is born, only to threaten millions in the future? For in spite of all the claims of geneticists for a glowing future for 'genetical engineering', there is a long way to go before they will be able to right faulty genes. Is it not better to prevent their increasing in the first place? Nature herself ensures that the majority do not survive, for 20 per cent of all conceptions, in the Western countries at least, die before term, and many others die at or just after birth through pre-natal causes.

Suppose medicine could save them? Many would still be condemned, condemned to a life totally dependent on drugs or mechanical aids. One such condition is phenylketonuria which is caused by a recessive gene and affects one in every thousand babies born in the United Kingdom. In this condition the child lacks an enzyme called phenylalanine hydroxylase which is made in the liver. This substance converts phenylalanine, which is found in proteins, into another chemical known as tyrosine, and without this reaction the phenylalanine builds up in the blood-stream to toxic levels and affects the brain, resulting in mental retardation leading to imbecility. Babies are now screened at birth for this condition and the treatment is to put them on low-phenylalanine diets as soon as possible, certainly before the age of three months. This diet must be rigorously enforced throughout life, otherwise these people would lapse into imbecility. The mental strain on the parents of a child born with such a defect must be tremendous, knowing as they do that one small lapse of their vigil in maintaining such a rigorous regime will send their 'healthy' child into idiocy. Is the strain imposed on both parents and child justifiable? Perhaps only the parents can answer.

But this is just one condition—there are many transmittable

genetic disorders that are being passed on into greater numbers of people. One that is very well known is haemophilia, a condition in which the sufferer can bleed to death if he cuts himself. This also is caused by the lack of an enzyme which in a normal person promotes blood clotting.

Apart from the genetical inheritance we are passing on to future generations, there is also the burden of treating and caring for those with dreadful handicaps which the sufferers, their families and the community have to live with. One of these conditions is mongolism; one out of every three mongols is born with heart defects and would not naturally survive. But because of the skill of the modern surgeon in correcting the heart defects, the number of mongols surviving is increasing and has more than quadrupled in the past thirty years in Britain. The question must be asked whether the saving of such people can be justified. Ethics supposes that every human being that survives birth, at least, has a right to live, and the doctor—and perhaps the parents—are thus faced with a dilemma which only their consciences can answer. Should decisions of life or death for the genetically impaired infant be left to society? The community in many cases will have to look after these unfortunate victims, and will have to supply the funds to keep an ever-increasing number of them in the future. In cases where an inheritable genetical defect is detected early enough, say before or just after birth, have we a right to save the individual at the expense of future generations? Even in cases where a child may live a comparatively normal life, should it be allowed ultimately to reproduce and so ensure the continuation of the fault into the generations that follow? Are we not aiding and abetting a fifth column which will bring down our own species? In a world which condones slaughter and torture, it seems hypocritical to be horrified at such questions when the survival of our species may be at stake.

Genetical pollution is very hard to detect except in the cases where there is a deformity or other obvious symptoms. But it goes further than these: the tendency for individuals with low intelligence to have large families and those with high IQ's to have small families perpetuates mediocrity and ensures a dilution of ability.

Society must decide on the solution of these problems, but, as with all the difficulties we find ourselves in, the longer we postpone making decisions the worse the problems will become.

There are millions of people alive today who in earlier times, while they would probably have survived birth and early infancy, would never have reached maturity. These are the people with inadequate personalities who would have been unable to cope with the battle for survival in a natural environment. Modern medicine and the cosseting by welfare states have ensured that millions of such people survive today, but, despite this, some are still unable to face the realities of modern life and seek to escape by indulging in extremes of self-pollution.

Not content with polluting his environment, Man deliberately pollutes himself. The more usual pollutants are tobacco, alcohol and drugs of one kind or another. A drug has been described as 'any chemical substance that alters mood, perception or consciousness and is mis-used to the detriment of Society'. Some of these substances produce in the user a dependence so strong that he cannot face life without them and becomes addicted. Addiction is extremely difficult to break, and more often than not leads to the degeneration of both body and mind accompanied by social degradation and ostracism; ultimately it can result in death. Many addicts are people with severe personality disorders. One psychiatrist has gone so far as to say that in one sense addiction is a good thing because, even if it does not lead to the death of the addict, it prevents his breeding and passing on his weakness. Alcohol, morphine, heroin and the barbiturates are all addictive drugs. Smokers are, in varying degrees, addicts too. Tobacco smoking does not have the effects on the mind and personality that alcohol and drugs have, but nevertheless smokers do expose themselves to the risk of bad health and premature death. Cigarette smoking, which is a deeply-rooted habit if not actual addiction, is a major factor in lung cancer, bronchitis and coronary diseases. In the United Kingdom a fight for and against tobacco has been raging for some years; the Health Education Council spends about £100,000 annually on publicity trying to dissuade people from smoking, while the big tobacco companies spend something in the region of £18 million on encouraging people to keep up with their self-pollution. The Royal College of Physicians published a report on tobacco smoking in 1962 which caused a ripple of concern and a temporary drop in the use of tobacco; some local authorities opened anti-smoking clinics, but, as the scare died down, one by one these were closed. Today there are less than a dozen, and only one of these is permanent.

The Treasury derives more than £1,000 million from tobacco duty annually, but in terms of human suffering tobacco is directly or indirectly responsible for the loss of 150,000 working man years per year and a death toll of 70,000 a year.

The drinking of alcohol is almost as old as Man himself and this probably accounts for the tolerant attitude we hold towards the drinker—at least until he gets drunk. The man holding his pint of beer somehow conjures up an air of masculinity, but when later he is seen rolling about the street he presents a sorry sight and we think he should be locked away. Like most things, alcohol can be taken in moderation without apparent ill-effect, but it affects people in different ways; some can take more than others and their efficiency is less impaired. Some individuals develop tolerance to alcohol during their lives and may never reach the extreme condition of alcoholism, but sometimes this creeps up on them and the brain chemistry is altered. The cells of the brain get used to functioning in the higher concentration of alcohol available and, if it is removed, breakdown occurs, accompanied by terrifying withdrawal symptoms. The chronic alcoholic, if not treated, falls victim to agonizing fears due to the effects on the brain and eventually suffers a complete mental breakdown; ultimately he has an uncontrollable nervous twitching of the body, *delirium tremens*. As well as damaging the brain, excessive drinking damages the liver and the blood circulation and leads to obesity. Chronic alcoholism leads to cirrhosis of the liver, malnutrition, peripheral neuritis, alcoholic epilepsy, psychosis and premature death. Many alcoholics commit suicide. In a recent Scandinavian survey of 220 male alcoholics it was found that fifteen killed themselves within the first five years of their discharge from hospital after treatment. Some alcoholics will not accept treatment, or, if they do, they soon relapse after their release. They will always be a burden on the community.

The social consequences of alcoholism are severe. It can cause tragedy through road accidents and tragedy to the families of alcoholics, for the alcoholic frequently neglects them. Absenteeism, unemployment leading to debt and crime are integral parts of the problem. There are 100,000 acute alcoholics in England and Wales, and the upkeep of those who cannot work, or who end up in prison, and of their families costs the taxpayer £6 million a year. There are three times as many who, while not in the same dire condition as acute alcoholics, are habitual absentees and poor workers. Their 'indulgence' costs the country another £35 million a year. Many alcoholics cut themselves off from any company except that of other alcoholics, and these tend to gather together in areas known as 'skid rows' where they can find anonymity and can indulge in their drinking excesses without stricture. Skid rows present a grave problem to public health authorities because they act as reservoirs of sick people and have a magnetic attraction for others, who are eventually drawn into the net. Many alcoholics die lonely deaths. Many others get themselves arrested for being drunk and disorderly because they regard the gaols as home, a refuge where they can obtain food and shelter. The National Council on Alcoholism was set up in 1962 with the aim of providing information centres to inform the public of the dangers, but alcoholism is a very difficult condition to treat because of the numbers involved, the inadequacy of the facilities and the resistance of many alcoholics to treatment. In Britain alcohol is the fourth largest cause of premature death, while in the United States it is next to heart disease and cancer as a killer. Although alcohol is a drug, far more publicity is given nowadays to the taking of opiates (the commonest is heroin), barbiturates, amphetamines, 'weed' (cannabis, marihuana) and hallucinogens (for example, LSD). We have got used to alcohol, and our attention is now turned to these other drugs, the use of which is rocketing, particularly among the young where the publicity has helped to increase their attractiveness. It is now the 'in' thing for young people to experiment with drugs, and many are led into drug-taking to enable them to be accepted among their contemporaries. Drugs also offer a new way of expressing revolt against the 'Establishment', more often than not represented by their parents. Among more mature people there is a dependence on drugs which is aided and abetted, although he may be unaware of it, by the family doctor who in prescribing drugs for depression and anxiety may be providing the agent for this dependence. Barbiturates, for example, are prescribed in large quantities as sedatives; it has been estimated that in Britain enough of these are distributed each year to provide twenty tablets per head for the entire population. Their action is similar to that of alcohol, and in society's tolerance of them they are rapidly reaching the same position as alcohol. Society tolerates far more the abuse, addiction and suicide that the use of these drugs brings than it does any of the others. Barbiturates are the most commonly used addictive drug.

All drugs are dangerous to health if not taken under controlled conditions. The ones that the addict takes to stimulate him or increase his pleasure, to separate him from the reality of life or even to promote hallucinations, are dangerous to his mind because he becomes so dependent on them that he cannot give them up. Like alcohol, drugs lead to progressive mental, moral and physical deterioration.

Marihuana, or cannabis resin, or hashish, the 'weed', which is usually smoked in cigarettes or 'reefers', is one of the drugs used to escape from reality and can give hallucinations which exaggerate experience. Marihuana turns a man in upon himself, and in the Middle East it is as acceptable in society as alcohol is in the West. There is reason to believe too that it is a car-cinogen (cancer producing) and that it will cause genetic damage to the children and even grandchildren of addicts. The burning of marihuana, as in the reefer, produces chemicals which are known to produce cancers in animals, and those components which pro-vide the 'kick' cause damage in the genetical apparatus of the cells.

Of the hallucinogens, LSD (LSD-25, d-lysergic acid di-ethylamide) is the most popular and the most powerful. Micro-gramme amounts can have significant effects in producing vivid hallucinations not unlike those that occur in psychotic patients. These hallucinations can be disastrous. There is a record of one man on an LSD 'trip' who stepped out of a third-floor window brimming with confidence that he could walk on air—he couldn't of course! In his book *Drugs* (1967), Peter Laurie sums up the LSD experience very neatly: 'The drug dissolves the crust that separates us both from the sensually experienced world and our own unconscious. The effect on civilised man is often that he discovers—to his surprise—as large and strange a world inside his head as there is outside.'

Amphetamines and caffeine (the latter found in tea and coffee) are stimulants, blotting out anxiety and releasing the forces of confidence and euphoria. The amphetamines, for instance, in-crease awareness and help people to keep awake. They have also been known to cause hallucinations and to produce irresponsible behaviour in the takers. Cocaine is a very fast-acting stimulant increasing, as long as its action lasts, all forms of physical pleasure. The reaction to the drug is severe and it is very difficult to break addiction to it. The opiate heroin is particularly popu-lar—it is all too easy to become addicted to it and dependence soon reaches an absolute stage. In the United States five people die every day of heroin poisoning; three of them are under eighteen and three of them die in New York City. In New York State there are more deaths from heroin than from any other single cause in the fifteen to thirty-five age group. Between 1960 and 1966 overdoses of heroin killed between 200 and 380 persons a year in New York City. In 1967 the number rose to 655, in 1968 to 730 and in 1969 to 950—254 of the 1969 total were under eighteen. The total number of heroin addicts in the United States is believed to be between 150,000 and 180,000, and Dr B. Louria of the Department of Public Health and Preventive Medicine at New Jersey College of Medicine and Dentistry in Newark estimates that the total number of addicts in New York State alone is between 60,000 and 100,000. In 1961 there were only a handful of heroin addicts in Great Britain; today there are about 2,000.

The opiate addict is driven on from one 'fix' to the next by his fear of the withdrawal effect which will ensue if he cannot obtain his supply of the drug. Withdrawal from opiates is the ultimate degradation in human behaviour. Dr Robert de Ropp in *Drugs and the Mind* gives a vivid description of heroin or morphine withdrawal:

> About twelve hours after the last dose of morphine or heroin the addict begins to grow uneasy. A sense of weakness overcomes him, he yawns, shivers, and sweats all at the same time while a watery discharge pours from the eyes and inside the nose which he compares to 'hot water running up into the mouth'. For a few hours he falls into an abnormal tossing, restless sleep known among addicts as the 'yen sleep'. On awakening, eighteen to twenty-four hours after his last dose of the drug, the addict begins to enter the lower depths of his personal hell. The yawning may be so violent as to dislocate the jaw, watery mucus pours from the nose and copious tears from the eyes. The pupils are widely dilated, the hair of the skin stands up and the skin itself is cold and shows that typical gooseflesh which in the parlance of the addict is called 'cold turkey', a name also applied to the treatment of addiction by means of abrupt withdrawal.
>
> Now to add further to the addict's miseries his bowels begin to act with fantastic violence; great waves of contraction pass over the walls of the stomach, causing explosive vomiting, the vomit being frequently stained with blood. So extreme are the contractions of the intestines that the surface of the abdomen appears corrugated

and knotted as if a tangle of snakes were fighting beneath the skin. The abdominal pain is severe and rapidly increases. Constant purging takes place and as many as sixty large watery stools may be passed in a day.

Thirty-six hours after his last dose of the drug the addict presents a truly dreadful spectacle. In a desperate effort to gain comfort from the chills that rack his body he covers himself with every blanket he can find. His whole body is shaken by twitchings and his feet kick involuntarily, the origin of the addict's term, 'Kicking the Habit'.

Throughout this period of the withdrawal the unfortunate addict obtains neither sleep nor rest. His painful muscular cramps keep him ceaselessly tossing on his bed. Now he rises and walks about. Now he lies down on the floor. Unless he is an exceptionally stoical individual (few addicts are, for stoics do not normally indulge in opiates) he fills the air with cries of misery. The quantity of water secretion from the eyes and nose is enormous, the amount of fluid expelled from stomach and intestines unbelievable. Profuse sweating alone is enough to keep bedding and mattress soaked. Filthy, unshaven, dishevelled, befouled with his own vomit and faeces, the addict at this stage presents an almost subhuman appearance. As he neither eats nor drinks he rapidly becomes emaciated and may lose as much as ten pounds in twenty-four hours. His weakness may become so great that he literally cannot raise his head. No wonder many physicians fear for the very lives of their patients at this stage and give them an injection of the drug which almost at once removes the dreadful symptoms. If no additional drug is given the symptoms begin to subside of themselves by the sixth or seventh day, but the patient is left desperately weak, nervous, restless and often suffers from stubborn colitis.

Hardly the behaviour one would expect from the most 'successful' animal on this planet!

Tragedy may also result when drugs are used in the practice of medicine. One of the greatest of these in the post-war period was caused by the use of the drug thalidomide, which was responsible for thousands of malformed children. Thalidomide was used in several countries as a sedative and particularly for women in the early stages of pregnancy. In 1961 the first hint that the drug might be having harmful effects on the developing human embryo was reported. Large numbers of children with congenital malformations began to attend clinics in Germany. The deformities were usually associated with damage that would have occurred between the 28th and 42nd day after fertilization. They included reduction deformities of the limbs, malformations of the eyes and ears, defects of the heart and kidneys and deformities of the alimentary canal. Something like 2,000 to 3,000 babies were affected in Germany alone. I do not want to labour this point because the victims of this tragedy are now growing up, and both they and their parents must find it painful to be constantly reminded of their misfortune. Whether the drug itself caused the defects, or whether it saved an already deformed foetus from being naturally aborted, has not been determined. The fact is that, if sufficient research had been carried out on the substance before it was used in human beings, this tragedy would probably not have happened. These otherwise normal children—there was no damage to the central nervous system—with all their natural desires have to bear their frustrations, but we will have them on our consciences for a long time because of the negligence of our science. Teratogenicity, or deformities in the embryo and foetus, is well known in nature and normally results in abortion, but the action of drugs may help to bring them to full term and so increase human suffering in the world. On the other hand, drugs may cause malformation and so increase the incidence of teratogenicity.

One of the greatest leaps forward in the conquest of disease was the discovery that the excretion of a mould, *Penicillium notatum*, would inhibit the growth of certain pathogenic (disease-producing) micro-organisms. Since that initial momentous discovery a whole range of antibiotics has been found and developed, and it is this armoury that has for all intents and purposes eradicated many of the diseases that once kept mankind's numbers in check. But there is a very severe drawback in reliance on antibiotics—the ability of living organisms, from the greatest to the smallest, to adapt to new situations. Among the millions of viruses and bacteria there will be some that have an immunity to the antibiotic or drug. These will survive a chemical and antibiotic attack and reproduce and eventually fill the niche left by their less successful relatives. A resurgence of a number of those diseases caused by drug-resistant organisms has already occurred, and unless our medical sciences can keep a step ahead we are going to be in a far worse situation than we were before we had the drugs. Our forbears at least had some natural immunity from these organisms, but our generation has been weakened by that very protection which was pumped into us early in life. Defended from disease through hygiene, inoculations and the like, we are left without natural defences to these new strains of pathogens. It is possible that before very long we shall experience new epidemics which would so strain the health service that it would be near to a state of total collapse. Even now, when a new influenza virus hits, the emergency bed schemes have to be brought in to deal with the sudden influx of sick. Doctors under pressure from the increasing numbers of patients in the normal day to day business of their profession can cope only by liberally dispensing antibiotics.

The discovery of drug-resistant micro-organisms in Man is paralleled by a similar discovery in animals. In the case of animals this situation has arisen out of the practice of adding antibacterial drugs to animal foods and from the oral administration of antibiotics. The latter was introduced after it was discovered that if animals are fed with small amounts of antibiotics they increase their growth rate. Although it is not known how these 'doctored' foods produce their growth-promoting effect, it is thought that this may be linked to a reduction in the alimentary flora. For nearly two decades pig and poultry farmers in Britain have been feeding their animals on diets containing tetracyclins and penicillin. Other substances have also been used in more recent years, and most of the other anti-bacterial drugs that are commercially available have been used from time to time to prevent outbreaks of disease. In the United States a larger number of drugs are used in animal foods for nutritional purposes than is permitted in Britain; also a greater number of domestic species are given these drugs.

As with humans, the emergence of resistant strains of pathogenic organisms in animals may mean that the animals will not be able to resist the new strains, and new drugs must be found to combat these. The new drugs, if not controlled, would lead to further modifications of the bacteria and the situation would be repeated. Medicine must keep one step ahead all the time. If there was a significant time gap between the identification of a new pathogen and the development of the drug to combat it, it could prove disastrous from a food production point of view. There is also the risk of cross-infection between humans and animals. A drug-resistant strain of a pathogen in an animal could be passed on to humans through their eating animals dosed with antibiotics. Already, resistant strains of disease-producing organisms are causing a large proportion of livestock diseases and there is a very real fear that human health may be affected. We just do not know enough about the effects of antibiotics in either the short or the long term. They may be throwing the body's metabolism out of gear, which over the long term may alter the physiology and genes of animals and Man; their cumulative effects have not been studied in anything like enough detail.

We may pat ourselves on the back that, if we don't smoke, or drink, or take drugs (putting aside the occasional medicinal lapses), we are not guilty of self-pollution, but how many people realize that in our food we take in countless numbers of foreign substances day after day, month after month, year after year? Apart from the antibiotic residues in meat, the pesticide and fungicide residues in our fruit and vegetables, our food contains what are called 'additives'. Today it is almost impossible to purchase food which is not polluted in this way and, although we may not be aware of it, it is our own fault. The modern way of life demands that less and less time is given to the preparation of food from basic raw materials in the home. We demand convenience foods, pre-prepared, pre-packed, on demand, out-of-season. Such a demand can be satisfied only by the treatment of food to preserve it—we should get an awful shock if we knew how old some foods are when we buy them. It is not uncommon for there to be more than ten additives in a single food product —combine a few of these to produce an instant meal, and the number of additives could be quite startling. Take bread, for instance. First the wheat: the seeds may be treated with a minimal fungicide to avoid rot, the plants may be dosed with an insecticide, the harvested grain may receive a fumigant such as carbon tetrachloride or carbon bisulphide to kill any remaining insects, and finally a protectant such as methoxychlor. Next the flour: this is bleached (chlorine or nitrosyl chloride are common) and enriched (thiamin, riboflavin, niacin and iron to guard against vitamin deficiencies). Baking powder and salt have an anti-caking agent such as calcium silicate, and the starch may have been bleached with potassium permanganate. On top of these there may be sodium diacetate (mould inhibitor), monoglyceride (emulsifier), potassium bromate (maturing agent), aluminium phosphate (improver) and calcium phosphate monobasic (dough conditioner). Is there any wonder that modern bread looks and tastes like chewed-up blotting paper?

Much of our modern food is completely tasteless—the additives and processing have seen to that. This has been put forward as one reason why so many people, in the 'affluent' countries anyway, are overweight. To get the real taste of the food, or what it is supposed to represent, you have to eat far more of it than you would of the unadulterated real thing. But the real danger of these additives is what they are doing to our insides. How does the body cope with the glut of chemicals pumped into it day after day? How much of a strain is it on the organs to process them and dispose of them? Are they disposed of, or do they accumulate or turn into more dangerous substances? The complexity of the problem is immense and seemingly impossible to solve. The American Food and Drugs Administration, which is only required to answer whether the chemicals are in themselves poisonous or not, has thrown in the towel and admitted that it just cannot tell if all the chemicals in food even when used one at a time are poisonous or not. The manufacturers counter any criticism by saying that the amounts introduced are so minute that they could not possibly have any effect. But, and a big but, they do not say that these infinitesimal amounts are being ingested from hundreds of sources, day after day and year after year. Estimates of the long-term effects of these chemicals both by themselves and in combination with others have rarely been attempted.

Occasionally some deleterious effects emerge, but usually not until the product has been well and truly ingested by large numbers of people. Cyclamates (artificial sweeteners) are a case in point. They were used for years until somebody discovered that they could be poisonous. Before a product is put on the market it should be thoroughly investigated for both its short- and long-term effects.

We cannot be sure that the antibiotics, drugs, food additives and all the other things which we absorb into our systems are not doing us irreparable harm. The human being has a great potential for survival, almost from the time it is conceived right until old age, and is extremely adaptable, and superficially the human race seems to be thriving on its artificial aids. But the fact is that underlying this veneer there are many anxieties which show that things are not what they should be. These are problems not just for the sociologist and psychologist but ones which each of us must ponder. They are the result of our increasing separateness from Nature and the realities of our being. Within this framework we live and it is here that the problems of self-pollution are at their most striking.

CHAPTER 13
The megakillers

Pollution is a deliberate policy in war. The aim of warfare is to bring the enemy to his knees by destroying his military machine, economy and morale. Modern warfare concentrates more on the last two of these objectives; the destruction of the enemy's economy and the demoralization of his people will compel his armed forces to give up the struggle. To achieve this, the modern military machine is geared to destroy the industrial and breeding centres of the enemy, for here belligerents can have the greatest effects on each other. The consequences of this policy are that cities and towns can be reduced to rubble and conditions made so hard for the populace that their will to survive becomes personal and not national. Invading armies are then welcomed—the citadel gates are flung wide, the conquering armies have food and in their own interests help the community to rebuild—but the waste of energy and materials has been immense.

At the end of the Second World War the entire aspect of warfare changed when American B-29 bombers took off from the Pacific island of Tinian bound for Japan. Their mission, destruction; their weapon, the atomic bomb. The horror of the devastation of Hiroshima, shortly to be followed by that of Nagasaki, ushered in a new age—pollution was no longer confined to the warring countries. Since then, military strategy has been built on the possession of nuclear or thermonuclear devices, and the world has lived in constant fear of being blasted out of existence. The continuing brinkmanship of the political camps makes this more than ever likely in the 1970s, '80s and '90s, and the proliferation of nuclear weapons increases the probability of someone using them in anger. Full-scale nuclear war would result in total annihilation almost at once in the northern hemisphere, followed a little more slowly in the southern, a situation awesomely described by Nevil Shute in his novel *On the Beach*. The limited use of atomic devices is feasible but unlikely, for once these weapons had been unleashed escalation would be rapid. But even if by some fantastic strength of will the leaders of warring powers limited the size of the weapon used, the radioactive fall-out would probably so contaminate the atmosphere that the fate of mankind would be virtually sealed.

Even without these consequences nuclear warfare would be wasteful in these days of over-population, land scarcity and diminishing resources. The 'ideal' would be to destroy the enemy's population without destroying his lands, which would then be there for the taking. This aim would more easily be achieved by using chemical or biological methods of destruction, and needless to say 'inventive' man has wasted no time in exploring this avenue. The arsenal is swollen with a vast array of super-virulent organisms nurtured in the secret laboratories of the military bacteriologist and virologist. The chemist has added his list of the deadliest substances made by man, compounds so lethal that literally a drop can kill within seconds. It is a polluted world that can produce brains that are willing to devote their talents to making these terrifying weapons, probably the ultimate in human degradation.

The military machine acts mindlessly. When the political machine fails (and our political history is a story of failures) we rely upon the military machine to sort out the mess. But, unfortunately, military solutions have rarely achieved the aims for which they were designed. The inflexible military mind is unable to detect changes in situations—it is programmed to react to certain kinds of stimuli and all things appear black and white with no grey in between. Thought processes seem to be absent, expediency is the criterion as long as the organization is true to itself. To be efficient it has been trained to react swiftly in order to survive, and although it may be run down in peace time, its nerves are still taut, for it has to survive the enemies from within the community it seeks to defend while remaining on the alert in sufficient strength to combat an aggressor from without. Backed by scientific and technological servants, it is under constant pressure to keep ahead of its possible opponents. It has no time, even if it was capable of it, to consider the wider issues. But, like an over-specialized organism, it is doomed to extinction, and in the twentieth century it is hanging precariously on to life only by persuading the political animals that it is essential for their survival. The result of this is that the world goes on making weapons and materials which are deliberately designed to kill and destroy, to pollute the individual and the environment. This is 'pollute and be damned' on the grand scale, for the military are concerned only with achieving their objectives, and if this means the deliberate destruction of the environment then to them it is justifiable, a wonderful example of ends justifying means.

Today's arsenals of nuclear and thermonuclear weapons constitute a hazard wherever they may be. The U.S.A. has a stockpile of some 2,000 inter-continental ballistic missiles ready for launching from their protective holes in the ground or beneath the seas in submarines. The Russians are not far, if at all, behind the Americans. The destructive power locked in these rockets is of the order of one million megatons—about the equivalent of three hundred tons of high-explosives for everyone alive on the planet at this moment. One would have thought this a little excessive! Some time ago the American Defense Department in a clinical review of the situation estimated that if a nuclear war were to break out, the fatalities would run at 160 million Americans, 200 million Russians, with few, if any, survivors in Western Europe.

To keep ahead of each other the military have to devise new weapons or weapons of greater power, and these have to be tested.

The 'old-fashioned' uranium bomb that destroyed five square miles of Hiroshima, killing 70,000 Japanese on 6th August 1945, could not suffice for long, for only three days later a plutonium bomb was used to obliterate 40,000 citizens in Nagasaki. From then on the Americans started on a programme of nuclear tests, soon to be followed by the Russians and the British. In 1954 the first hydrogen bomb was exploded by the Americans over Bikini Atoll in the Pacific, unleashing a destructive power of the order of 20 million tons of TNT. The explosion was a thousand times more powerful than the 'primitive' bomb that destroyed Hiroshima, and it has been estimated that its power was greater than all the explosives used throughout the whole of the Second World War put together.

But although the immediate destruction caused by the explosion of a nuclear device is disastrous in itself, it is only part of the weapon's power. The radiation emitted contributes its share of disease and death, spreading its pollution far and wide. Radioactive substances, like all other pollutants, get into the life cycles of the environment and the fall-out from a nuclear explosion enters the atmosphere or the sea, where it may persist for years. Anyone exposed to radioactive fall-out can do nothing about it. The total effects cannot be estimated, but we have seen examples enough to be aware of the dangers. When the 1954 hydrogen bomb was exploded an area of the Pacific which was considered sufficient for the safe test of the weapon was cleared, but it was not sufficient for the Japanese fishermen on the *Fukuryu Maru* a hundred miles away. Twenty-three of them suffered varying degrees of radiation sickness and one died. The *Lucky Dragon*, as its name means in English, became the symbol of nuclear pollution. The Japanese fishermen were not the only sufferers from the military miscalculation: although hardly noticed by the world at large, the children on the island of Rongelap were to suffer too; about 80 per cent of them developed thyroid growths after the detonation.

And so the tests went on. Thousands of children born since the beginning of nuclear testing are likely to die of cancer and leukaemia (over-production of white blood cells) induced by radiation from them. But the effects are not limited to the generation living at the time of the tests. Radiation increases the mutation rate in the living genes, and generations of organisms, including man, yet unborn can be crippled or destroyed as a result. An increased mutation rate could reduce the lifespan of the human race.

This is wholesale pollution of the environment in time of peace, yet in spite of all the warnings and the experience gained from the early American, Russian and British tests, the 1960s witnessed the entry of two more countries to the 'Nuclear Club', China and France, who still persist in exploding nuclear and thermonuclear devices in the atmosphere. What the Chinese (and Russian) tests have done to central Asia can only be guessed at, but the Asians at least will reap the benefit of their own nuclear aspirations. The French in testing devices over the Pacific are polluting the sea as well, but both groups are poisoning the atmosphere, the air we breathe. What will future generations think of countries which, in spite of warnings, still want to carry out programmes of nuclear testing?

But what of America and Russia? Although they have been persuaded from atmospheric testing, they haven't given up. They have tried desperately to continue and even increase their programmes of underground tests and to take their weapons out into space (see Chapter 11).

An act of total irresponsibility was carried out by the United States on 9th July 1962 when they exploded a nuclear device in the upper atmosphere. It caused widespread disruption of the ionosphere and blacked out long-range radio communications over much of the globe. A new radiation zone was created, which luckily was short-lived. To experiment with nuclear devices in the layers of the upper atmosphere, of which we know so little, is both stupid and criminal.

The military are completely blind to the wider implications of their activities. Why do they want yet bigger and better bangs? A lack of imagination prevents them from seeing that there could be trouble for us if they detonate nuclear bombs in near space. The explosive power and the radiation could distort the Earth's protective shields to such an extent that the radiation from the Sun and deep space could reach the Earth's surface. If this were to happen it could destroy life on this planet; even with a slight increase in radiation the human organism could be sterilized or even wiped out.

But even in situations supposedly under control, accidents will happen. Rockets may explode accidentally and nuclear bombers may crash. We know in fact that two bombers have crashed, one near Thule in north-west Greenland and one near Palomares in southern Spain. The latter is a very good example of military authorities' insensitivity towards ordinary people who, through no fault of their own, can become involved.

On 17th January 1966 an American Air Force nuclear bomber carrying four hydrogen bombs collided with an air tanker which was trying to refuel it over the villages of Cueras and Palomares in Spain. The entangled planes crashed and parts of the wreckage and bodies fell in the village of Palomares. Not far away were two of the hydrogen bombs which had floated safely down in their parachutes, thanks to the prompt action of the crew, but unfortunately the explosive detonators had gone off, splitting the cases and releasing the plutonium. Although there was no chance of a nuclear explosion, contamination of the soil and surrounding vegetation was inevitable. Of the other two bombs, one landed safely near the coast and the other landed in the sea.

After the accident the insensitivity of the authorities, both American military and Spanish, showed itself. The Americans immediately launched 'Operation Broken Arrow', the Strategic Air Command's poetic code name for nuclear accidents. Under this procedure a cloak of secrecy descends upon the area of an incident. The villagers of Palomares were not told what had happened, but, if they had been, the Americans would have known much earlier where the fourth bomb had come down and also which villagers had been in contact with the spillage from the bombs. Little did the villagers know that the black stuff they had found coming from the bombs was plutonium which could render their soil unfit for cultivation possibly for up to 24,000 years. All they knew was that they were forbidden to go into their fields and they were subjected to constant medical checks. This may have been in the interests of the peasants' physical health, but it had little regard to the mental strain they were undergoing. Their village was covered with mystery and as a consequence was ostracized by its neighbours.

In the end the authorities were forced to let the villagers know that a large area round Palomares was contaminated with plutonium and uranium 235. A bit of American bombastic showmanship occurred when the American Ambassador to Madrid visited the scene, had a swim in the sea off Palomares, and is reported to have said, 'If this is radioactivity—I like it.' All right—if you have not been contaminated or lost your livelihood through it. In the circumstances, a more insensitive remark could not be imagined.

The Americans removed 1,750 tons of contaminated soil and shipped it to the South Carolina nuclear dump. The lower active soil was ploughed under, subsequently given a clean bill of health and given back to the peasants, but, according to some reports, cattle still died. The very next day, it is reported, a farmer lost one of his best productive cows, and although he was promised reimbursement if he kept quiet about the incident, this promise was never kept.

It took eighty days of searching to locate the fourth bomb in the sea and a million dollars to recover it. If it had not been found, but instead had been carried far by the undersea currents and had broken open on some rocky shore, the result would have been very nasty.

The fight against the high-handed treatment of the villagers of Palomares was led by a courageous lady, Luisa Isabel, Duchess of Medina Sidonia. Her battles on their behalf and on behalf of the fishermen of the Andalusian village of Sanlucar against the dumping of radioactive wastes, including plutonium, in the sea from the American Air Force base at Rota are now well known. If the military authorities had been sensitive to people's feelings in the first place much of the damage could have been prevented.

Earlier I mentioned briefly that, not content with developing bigger and better bombs, the modern exponents of warfare have reinforced their arsenals with the equally deadly chemical and biological weapons. We tend to think that these are the products of twentieth-century technology, but in fact they have been used for centuries, albeit with an effectiveness minute compared with the capabilities of the modern versions.

In 600 B.C. Athenian troops poisoned their foes' drinking water, and in the Peloponnesian War sulphur dioxide was used in the siege of Plataea. In the First World War 100,000 persons on both sides were killed by some 25,000 tons of the poisonous gases chlorine, phosgene and mustard. The use of these agents has always been emotive, and after the First World War feelings ran so high that an international effort was made to impose some form of control. In 1925 thirty-eight nations signed the 'Geneva Protocol for the Prohibition of the Use in War of Asphyxiating, Poisonous or Other Gases, and of Bacteriological Methods of Warfare'. Although much of the initiative came from the United States, it is ironic that that country has never ratified the Protocol, although it has constantly affirmed that it will never resort to the use of chemical weapons unless they are first used by its enemies. Eighty-four nations are now parties to the Protocol; these include all NATO members except the U.S.A., all members of the Warsaw Pact including the U.S.S.R., all nuclear powers (apart from the U.S.A.) including the People's Republic of China, and all major industrial nations except Japan and of course the U.S.A.

The public squeamishness which the idea of chemical and biological warfare generated certainly prevented their use in the Second World War, with the exception of Japan who used gas in China. But this squeamishness does not always extend to the use of harassing agents such as tear gas or CS gas, and defoliants. Whether or not the Geneva Protocol covers these agents has always been disputed. The United States has maintained that it does not, and in Vietnam the Americans have made widespread use of them, with tragic results. Many countries use tear gas and CS gas for riot control as a matter of course.

CS (orthochlorobenzalmalonitryl) gas, developed by the British (hardly a matter of pride), is an 'improved' poisonous tear gas which only temporarily incapacitates the victim, but it can do this in twenty seconds. It causes difficulty in breathing, a tearing painful cough and, if in sufficient quantity, nausea and vomiting. The chemists cannot tell whether it has any lasting effects—probably not, but it is the principle which is important.

The gas 2, 4, 5-trichlorophenoxyacetic acid, 2, 4, 5-T, was used by the Americans to destroy the heavy foliage over hundreds of square miles of the Vietnamese jungle, the idea being to clear the cover used by the Vietcong. Another gas, 2, 4-dichlorophenoxyacetic acid, 2, 4–D, has also been used and it is claimed that this one has a short life in the soil and relatively low levels of toxicity in man. But the overall effect of the use of these defoliants in Vietnam has been to cause severe ecological damage. The poisoning of crops and the persistence of 2, 4, 5-T in the soil will render the Vietnamese dependent on aid from outside, and it is a doubtful victory for capitalism if South-East Asia can be saved from Communism only at the expense of reducing the people to beggary. The defoliation campaign has already had a deep psychological impact on the Vietnamese and has markedly affected their attitude towards the Americans.

The Geneva Protocol has another and very serious loophole: it does not prohibit the research, development, testing or production of chemical and biological weapons. In the case of chemical agents in particular, thousands of tons were available during the Second World War, nerve gases hundreds of times more poisonous than those used in the First World War. Since then the United States and, it is believed, the U.S.S.R. have been stockpiling these gases to an alarming extent and 'improving' on them. The nerve gases act directly on the nervous system, producing paralysis and poisoning; the most common ones are G.B., isopropylmethylphosphonofluoridate or 'Sarin', produced originally by the Germans during the Second World War, and VX, the chemical formula of which is still secret. Sarin is a volatile liquid which evaporates to a colourless and odourless gas and attacks the victim's respiratory system. If the concentration of Sarin in the air were ten milligrammes per cubic metre, a lethal dose would be accumulated in a matter of ten minutes. VX is also a liquid, but is several times more toxic than Sarin and much less volatile. It is lethal when inhaled or merely deposited on the skin and kills in a few minutes. Further, by contaminating the ground it can make an entire area hazardous for many days. So efficient are these gases that it would require less than fifty tons to wipe humanity out. Until recently the huge arsenals of the United States and Russia were producing sufficient quantities to wipe out the world's population several times over, which seems an utterly pointless activity, for you can only kill a person once.

A more subtle approach may be made with the use of hallucinogens added to an enemy's water supply or distributed as dust. Lysergic acid diethylamide (LSD) would be ideal, for example: two pounds of this substance would affect up to 10 million people. Depending on the amount ingested, the effects would range from complete madness to slight hallucinations. The breakdown of society caused by such an attack would be very favourable to the enemy, but once hallucinogens were in the environment it would be very difficult to limit them to the country under attack because they would be carried in the atmosphere and water circulation for some time.

But most to be feared are the nerve gases and the biological weapons. The possibility of biological warfare is perhaps the most despicable of all. Not only can the biologist, by selective breeding, produce super-virulent organisms, he can produce them in great quantities and store them over long periods. They can be distributed in aerosols and some of them, if they could be evenly distributed all over the world, would wipe out the human population. For instance, a pound of one organism, *Botulinum toxin 1A*, could kill everybody on the planet. If used in small amounts over an enemy territory it would sweep the country clear of humans after a few days, and because it decomposes in the air after a few hours the conquering army could move in. Other bacteria, for example anthrax, *Bacillus anthracis*, are very resistant and can live in a reduced activity form over long periods, only to become active again when a suitable host animal picks them up. Off the west coast of Scotland there is a small island, Gruinard Island, which is still uninhabitable and will probably remain so for hundreds of years because of the anthrax that was put there in a proving test during the Second World War. Less than a millionth of a grain of *B. anthracis* from an aerosol is believed to create a 50 per cent chance of contracting pulmonary anthrax. Symptoms may first be put down to the common cold, but soon there would be severe coughing, cyanosis, respiratory failure and death. Untreated pulmonary anthrax is almost always fatal. The World Health Organisation Report on the effects of biological agents predicts that an aerosol attack of fifty kilogrammes of anthrax agent would cause a high mortality over at least twenty square kilometres, and some would regard this as a very conservative estimate.

Pneumonic plague, cholera and other organisms which can produce epidemics or pandemics can also be used by the biological warfare scientist, and their use could be particularly underhand. A few agents or diplomats could very easily smuggle into a country sufficient organisms to cause enough damage to crops, livestock and humans to create severe difficulties in that country at no risk of suspicion to the aggressor. Sudden outbreaks of crop diseases, animal diseases such as foot and mouth and human diseases which quickly reach epidemic proportions are not rare even in this day of hygiene. They are more likely to get a hold in the less developed countries from where, with the help of modern transport, they can spread very rapidly. Perhaps the only thing that might dissuade a would-be user is the fact that there may be a come-back if the disease got out of hand.

A similar state of affairs may be protecting us from an underhand attack with chemical agents. It would be quite easy for an aggressor to poison the food or water supply of a city and get away with it, though this is less subtle than the biological agents, and an immediate alert would be put out and steps taken to find the polluter and to protect other supplies; but not before some damage would have been done.

Whatever the agent, the hard fact remains that vast quantities of them are lying around and it cannot be expected that such quantities of chemical and biological weapons can be stored *ad infinitum* without some leakage. Several disquieting incidents during 1969 really made people sit up and take notice. A series of earthquakes which shook the city of Denver in the U.S.A. was traced to underground disposal of chemical wastes at the U.S. Army's Rocky Mountain Arsenal. VX nerve gas was accidentally released over a wider area than intended during a test over the Dugway Proving Ground in Utah in the United States, with the result that 6,000 sheep grazing forty miles away died. It was fortunate, but not for the sheep, that the wind was in their direction, for otherwise the poisonous cloud would have drifted over Salt Lake City; what might have happened then can be left to the imagination. Although the U.S. Army doled out a handsome compensation to the owners of the sheep, they had the nerve to disclaim all responsibility for quite some time. A further release of nerve gas occurred at an American base on Okinawa and more than twenty people had to be taken to hospital. Subsequently it was reported that America was also storing nerve gases in Germany, and these incidents put the dangers on a world scale.

Public interest and anger resulted in other shattering revelations. Chemical and biological weapons become obsolete and deteriorate with age, and ways have to be found of disposing of them. The choice is either chemical inactivation, which is complicated and expensive, or dumping in some convenient spot, which is relatively cheaper. Inevitably the cheapest solution is sought and that seemingly limitless volume of water called the ocean is regarded as a suitable dumping ground, without any heed to the possible consequences. The extent to which the sea has already been used in this way was only revealed when in the summer of 1969 some American Congressmen got wind of a plan to transport 27,000 tons of war gases, most of them nerve gases, across the United States to the coast for dumping in the North Atlantic. An inquiry was demanded and it was revealed that the gases stored in rockets and bombs were to be loaded into some 800 railway wagons at the Rocky Mountain Arsenal near Denver and moved through such cities as Indianapolis, Knoxville, Cincinnati and Philadelphia, and along the southern outskirts of metropolitan New York to the port of Earle, New Jersey. The gas was then to be loaded into old 'Liberty' ships, which would be taken some 125 miles out to sea and scuttled together with their deadly cargoes. The public outcry was tremendous, and rightly so. The army climbed down, but the incident brought to light the fact that dumping of obsolete weapons into the sea had gone on for some considerable time and that safety precautions had not been all that they should have been. Several questions remained unanswered; how long would the containers resist corrosion by the sea; what would happen if they ruptured either when dumped or later by corrosion; and if a container did rupture how wide an area could become contaminated? The polluters have given very little thought to these questions.

Despite the fuss, the American military bided their time, believing that people soon forget, and August 1970 saw them trying again, this time with more success. Some 66 tons of the original 27,000 tons consignment, contained in rockets encased in concrete coffins, were considered to be too dangerous for any means of disposal other than dumping them in the sea. Despite all efforts to stop them, the army in the full glare of publicity shipped the gas across America by another route and then out to sea, where the concrete coffins were laid to rest some 300 miles into the Atlantic. Britain, at the request of the Government of the Bahamas, who were concerned that any leakage might have severe effects on them, sent an official team of observers who, as expected, were reassured by the American military's explanation that if any leakage did occur the gases would be soon hydrolysed and rendered harmless. One would have thought that it would be far better in the first place to build hydrolysis plants and destroy the gases under controlled conditions than to place the world at risk. Any army that can put so much of its resources into manufacturing these weapons has a moral responsibility to provide for their safe disposal.

But America is not the only practitioner of dumping. Since the end of the Second World War, according to the Ministry of Defence, Britain has dumped some 200,000 tons of poison gas, including the deadly nerve gas tabun, into the sea, and some of it in the shallow sea at that. Some people believe that a leakage from this eleven years after the event was responsible for the deaths in the Irish Sea in 1969 of thousands of sea-birds, fish and seals. Some of the fish had peculiar scars on them reminiscent of gas burns and some seals were reported off Cornwall with similar burns. In the same year fishermen trawling in the Baltic fished up drums of gases that had been dumped there just after the end of the Second World War. They all received burns, two of them so severe that they had to have plastic surgery. The nets and fish had to be destroyed and the holiday beaches of southern Sweden and the island of Bornholm were evacuated.

It was the British also who created an earlier scare in the Caribbean area when, dismissing the fears of Cuba and Florida of a typhoid catastrophe, in 1954 we sent a floating laboratory with the intention of carrying out full-scale germ warfare experiments in the Bahamas. The exercise, Operation Cauldron, was regarded as purely defensive and Cuba was assured that in no conceivable circumstances could she be affected. The whole affair was clouded with secrecy, but *The Guardian* reported: 'Germs were not used. Their part was played by material similar to the weight of bacteria mixed with radioactive substances from the Harwell atomic piles.'

Although it is claimed that nerve gases hydrolyse in water, it is interesting to observe that evidence has been obtained to show that if a ship carrying a load of defoliants to Vietnam were to be sunk, it could totally poison the surrounding ocean for at least twenty-five years. In fact, the effects would be far worse than they are in the atmosphere since defoliants are not so easily dissipated in the sea.

The events of the last few years, tragic as they are or might have been, have at least led the American Government to modify its policy regarding chemical and biological warfare. The United States has renounced the first use of chemical weapons and will not use biological weapons under any circumstances, even in retaliation. It is expected that she will ratify the Geneva Protocol shortly, in spite of voting in 1969 against a resolution, passed by 80 votes to 3, in the United Nations General Assembly which declared that the Geneva Protocol does prohibit the use in war of all chemical agents directed at men, animals or plants.

All attempts to control the production of these and nuclear weapons must be encouraged to the utmost. The military objectives in any war could be met many times over by the arsenals now in existence. Unfortunately for the rest of the world the belligerents would not be the only ones to suffer even if a limited war, which is unlikely, were to take place. Those political leaders who advocate violence as a means of achieving their ends lack the imagination to see the wider implications of their policies. They also admit by their advocacy of violent solutions their incompetence as leaders. If the military are polluters then the politicians are the ultimate polluters.

CHAPTER 14
Dangerous dreams

There never was a time when Man did not dream of making the world a better place to live in. A 'Land fit for Heroes' was promised to the returning soldiers and sailors of the First World War. Their disillusionment was profound, and in England at least has affected the course of subsequent history. What seems to have been forgotten is that the world *was* once fit for men!

From the eighteenth century onwards advances in science and technology have led to a belief, amounting almost to faith, that Man could do anything he wanted to. All he had to do was to apply this new knowledge and he could create a new Eden. The shattering impact of, in Sir Winston Churchill's words, 'Perverse Science' in the Second World War made this technological realization nearer to Hell than Heaven. After the conflict was over some of the old dreams returned. Some of the schemes dreamed up by the men of the new technology could be every bit as, if not more, devastating as the war and, worse, their effects would last a long time after those of the war had been forgotten. Great ecological changes could be triggered off if only a few of the ideas were carried out. Some of the most dangerous of these dreams were associated with schemes to produce hydro-electric power and irrigation. The effects of diverting rivers and ocean currents are of such magnitude that few could imagine the wider aspects of their impact on the environment.

One of the earliest schemes was suggested during the American War of Independence: the idea was to block the Gulf Stream to deprive England of its warmth. But most 'practical' schemes have been concerned with the generation of power. One of these was a proposal to dam the English Channel across the Straits of Dover and use the water which flows from the North Sea into the Channel to drive turbines to produce electrical power. In addition it would form a rail and road link between England and France, so eliminating the need for a tunnel or bridge— if such a need exists! The scheme included a lock system to allow ships to pass through on their way to the Thames and the other North Sea ports. The cost was prohibitive and so the scheme never got under way, but nobody considered the ecological consequences the barrage would produce.

A more ambitious scheme was proposed by a Russian, Pyotr Borizov; he suggested that a dam should be built across the Bering Straits connecting the United States to the Soviet Union. If this were to be done, the Arctic would be transformed. The Russian engineers calculated that such a dam fitted with a series of massive pumps could draw huge quantities of cold water from the Arctic Ocean into the Pacific, so reversing the cold currents now flowing into the North Atlantic. The result would be that the Gulf Stream would carry its warm water right across the Arctic Basin into the Pacific—eventually melting the ice of the Polar Seas. The melting of the ice would only slightly raise the sea level, and the islands of Arctic Canada and vast stretches of Siberia would be warmer, and their rich mineral deposits more accessible and easily transported in the ice-free seas—or so the proponents of the scheme argued. What they missed was that the evaporation of the water would increase and consequently more cloud would be formed which would reduce world temperatures and plunge the northern lands into a pluvial age, an era of continual rain. The ecology of the North Atlantic, North Pacific and Arctic Ocean would certainly undergo irreversible changes with catastrophic effects on the fisheries, and although in certain areas harvests of cereals and other crops might have been increased in the short term, they would have been eventually destroyed.

A plan to dam the mighty Amazon has been put forward, which if it were carried out would create an enormous lake 70,000 square miles in extent. It would provide power for the growing cities of Rio de Janeiro, Montevideo and Buenos Aires, but what of the Indians living in the jungle of the Amazon basin? Most of them would be drowned or driven from their villages to starve to death. The stoppage of the water would affect the fisheries which are dependent on the silt brought down by the river. But perhaps the greatest loss, next to the lives of the Indians, would be the destruction of one of the world's last great expanses of tropical jungle, for millions of trees would be killed. Great changes in the moisture content and heat balance and, therefore, climate would be experienced over thousands of square miles of South and Central America.

In 1928 the German architect Herman Sörgel suggested a project which, if it had been implemented, would have done incalculable damage to the whole of the Mediterranean Sea area. His idea was to convert the Mediterranean into a vast lake by damming the Straits of Gibraltar and the Dardanelles. The result would be that the surface level of the Mediterranean would drop about three feet every year, thus providing a source of hydroelectric power at the mouths of the rivers flowing into it. Sörgel realized that the Mediterranean loses millions of cubic feet of water by evaporation every year which neither the rainfall in the region nor the rivers feeding the sea are sufficient to make up. At present water flowing through the Straits of Gibraltar from the Atlantic and water from the Black Sea make up this loss and Sörgel's idea was to shut off this supply and allow the evaporation to do the rest. At the end of about one hundred years, when the water level of the Mediterranean would be some 330 feet lower than it is today, it would then be possible to dam the Straits of Messina between Italy and Sicily and also the gap between Sicily and Tunisia, which would convert the Mediterranean into two huge lakes. From then on another drop of 330 feet in the water level would be possible well within a century.

The 'benefits' from Sörgel's project, which was called 'Atlantropa', would be colossal, or so he said. Most of the countries bordering the Mediterranean could claim huge areas of new land, as much as a quarter of a million square miles, including most of the Adriatic Sea. Venice would cease to sink, it would be high and dry hundreds of miles inland. Corsica and Sardinia and the Elba archipelago would once again be joined, and Tunisia, Sicily, Egypt and Spain would gain great areas of new land. The Costa Brava would be many miles inland and many of the Greek islands would form the high ground of the land claimed from the Aegean. Again the ecological consequences of this scheme were disregarded; most important, the circulation of the water and the mixing with the Atlantic would be stopped. The migration of fish, such as tuna, which today go into the Mediterranean to breed and form important food reserves for the region, would be halted. And although it is likely that compensating factors would establish themselves in the biological system, it is unlikely that they would be as productive as they are now. For a short while, as the nutrient salts increased through the combination of the inflow from the rivers and the evaporation of the water, a blooming of marine life might have occurred, but eventually the salinity would have increased, creating another Dead Sea. One thing Sörgel did not allow for in his calculations was the degree of pollution in the Mediterranean which would have turned what was left of this magnificent sea into a giant cess-pool. And what of the inhabitants of the new land if the dams burst?

Sörgel also suggested that if the Congo River were to be dammed near to its mouth, the Congo Basin would become a vast lake and the Lake Chad area could be converted into a great inland sea. This would have brought about colossal changes in the ecology and climate of the region.

Two other imaginative proposals for altering the Mediterranean area were the flooding of the Qattara Depression in North Africa and the damming of the River Jordan. A Swiss scientist, Dr Ball, had the bright idea of changing the Qattara

Depression into a lake. The Depression is an enormous expanse of barren sand dunes and salt plains nearly as big as Lake Ontario; over half of it lies 160 feet below sea level and near its southern end it is over 400 feet below sea level. There are very few fresh-water springs and oases in this vast area and only caravan tracks cross its barren interior. The idea in this scheme was to pipe or possibly channel water from the sea to fill the Depression and use the water coming in to generate electricity. Because of the high temperatures met with in the Depression, the evaporation would be enormous and produce rainfall, but even before this could happen, huge quantities of water would disappear into the dry ground. The only trouble, at the time the idea was put forward, was that there were no customers for the electricity! If this scheme had been carried out, it would have brought about great changes to the surrounding desert. Vegetation might have flourished and encouraged settlements on the new lakesides, but yet another foothold for water-borne diseases in North Africa would have been created. However, it may have been and still may be a means of making the desert flower, but before any such attempts are made the ecological consequences must be known. We have already seen that Man's sudden changes to the environment more often than not lead to trouble.

Although the damming of the River Jordan would not have such dramatic ecological effects as those of flooding the Qattara Depression, it would cause changes in the Sea of Galilee, the Jordan itself and the Dead Sea. Near its source the River Jordan flows through a marshy area called Lake Hula, just north of the Sea of Galilee. Between Lake Hula and Galilee the river drops over 600 feet, and from Lake Galilee it flows for nearly 200 miles before it empties into the Dead Sea. The Sea of Galilee is nearly 700 feet below the level of the Mediterranean and the Dead Sea is over 1,200 feet below that level, so the river runs from about Mediterranean sea level to well over 1,000 feet below in about 250 miles. The bed of the Dead Sea is over 2,500 feet below sea level. In the middle 1920s a Frenchman, Pierre Gandrillon, proposed a grand scheme for producing power in the Jordan valley. He saw the Dead Sea as one huge basin where the intense solar radiation evaporates the water at an enormous rate, leaving the characteristic salty Dead Sea—so salty that a swimmer cannot sink in it and nothing can live in it, hence the name. It is over seven times more salty than the oceans, with magnesium chloride as the most abundant salt, not sodium chloride as in the oceans. The average loss of water from the Dead Sea is over 3,500 cubic feet every second, which is made up by the water coming in from the rivers and in particular from the Jordan. It seems, however, that the rivers are slowly losing the battle, because the old shorelines that surround the area prove that the Dead Sea was much larger in the past. During the last twenty to thirty years the surface of the sea has dropped several feet, and this was important to Gandrillon's plan. His idea was to provide fresh water for irrigation and hydro-electric power for the region.

He proposed building a dam across the River Jordan, south of the Sea of Galilee, which would mean that all the water which normally runs down the Jordan valley would stay there and the enlarged inland sea so formed would be available for irrigation schemes in the surrounding country. And although the Jordan would disappear for a short while, it would reappear when the Sea of Galilee began to overflow the dam. Gandrillon then proposed that water from the Mediterranean should be pumped overland to a few miles below the Galilee dam and be used to produce power. First, the Mediterranean water would be allowed to build up in a large reservoir above the Jordan valley and then be allowed to drop in almost vertical pipes to the bottom, a distance of nearly 900 feet, to the turbines. These would be big enough to generate sufficient power for industrial and domestic use and also to pump water from the Mediterreanean to keep the head of water topped up. The water would then continue down the Jordan Valley into the Dead Sea, which would once again begin to increase in area because the evaporation would not be greater than the inflow of water. The plan was then to build a navigable canal which could carry freight from the Mediterranean, across the bed of the Jordan and down towards, but not quite into, the Dead Sea. Here another power station would be constructed to produce electricity from the water falling down the 600-feet drop to the sea. The electrical power produced at this power station would provide energy for a new chemical industry extracting and ultilizing the salts available in the Dead Sea. The surface area of the Dead Sea would increase, but so would the evaporation, so that, it was hoped, a new balance would be created. Another plan, similar to Gandrillon's, was suggested by an American, Dr Clay Lowdermilk, in the 1930s. If these plans were revised today they would need the co-operation of both the Israelis and the Jordanians, because up to the Six Day War in 1967 the Israeli border was well west of most of the Jordan valley; in the future it could well be a great venture for both countries. But danger remains inherent in all these schemes; increased evaporation, larger sheets of water and irrigation canals bring their own ecological problems.

Another idea that concerned the Middle East was suggested to get round the closure of the Suez Canal. The scheme was to cut another canal across the Sinai Peninsula from the Gulf of Aqaba to the Mediterranean swiftly by using nuclear devices— the dangers of nuclear fall-out just do not seem to be considered any more. The use of nuclear explosives has also been considered to blast an alternative to the Panama Canal across the Central American isthmus, and to dig a deep-water harbour at Cape Thompson in Alaska—the latter was abandoned after protests from the Eskimos about the radiation risks. The American Atomic Energy Commission (AEC) is still looking for work under its Project Plowshare—the peaceful uses of nuclear energy. Although the AEC lists a number of interesting projects that could be undertaken by using nuclear explosives, it does not provide estimates of the radioactivity that would be released by the detonations, nor does it consider the possibility of earthquakes being triggered off by the ground shocks generated by the explosions.

Even schemes which involve altering only local water masses need to be carefully worked out. Plans such as those suggested for Morecambe Bay, the Wash and the Severn and the Dee estuaries, could have considerable effects both on the land and in the surrounding waters. The engineering problems can be solved, economic and political differences in time smoothed out, but ecological effects are frequently irreversible and most often disastrous. After the engineering considerations have been taken into account, and the political and economic factors solved, the decision of whether or not a scheme should proceed ought to depend on ecological criteria.

We are living in an age where, backed by his powerful technology, Man can do or attempt to do almost anything he wants to. He can blast his way through mountains or cut through isthmuses with nuclear charges. He can turn rivers in their courses or deflect ocean currents. He looks beyond even these 'miracles' to a time when he can control the planet's climate, melt the polar ice-caps or create new planets out of old. The computer has provided a new confidence, a child-like faith that has given credence to his wildest dreams. His new tool, he believes, will enable him to predict all the variables his designs will bring about in the environment—dangerous dreams that could lead to catastrophe.

PART IV

PROJECT SURVIVAL

CHAPTER 15
'Twill do, 'twill serve

It is popular to write about pollution; it is the fashion to bemoan the excesses of the industrialized state. It is very easy to record the despoliation of our environment and to plot the mistakes of technology, but it is difficult to suggest practical ways of getting out of the dilemma. Yet we must not forget that a great deal has been done, albeit not nearly enough, to ameliorate the effects of one form or other of pollution and to get to grips with the problem. Already in this book we have examined some of these.

Some attempts to control the spoiling of the environment were made as early as the thirteenth century. The introduction of coal-burning was perhaps the turning point, for from then on the towns were never to be the same. So bad was the problem of smoke in London that in the year 1273 the use of coal was forbidden by law, and the reason given for this was that the smoke was considered 'prejudicial to health'. A short time later, in 1306, a Royal Proclamation prohibited the use of sea coal in furnaces. Despite this, the conditions gradually worsened and were in time accepted as a normal part of life.

The arrival of the Industrial Revolution intensified the problem, and living conditions became so bad that something had to be done if the citizens were to survive. The nineteenth century saw the first real attempts to control environmental pollution. A section dealing with factory smoke was included in the Town Improvement Act of 1847, and between 1853 and 1856 Smoke Abatement Acts were introduced to deal with the problem in

London. As the century wore on, more legislation was introduced to protect public health, the measures including further smoke control. The 1866 Sanitary Act, for instance, gave powers to the sanitary authorities to take action if smoke was a nuisance, and the 1875 Public Health Act contained a smoke abatement section. This particular clause formed the basis of the legislation that has continued up to the present day.

During this time public anxiety about smoke was acute and a number of voluntary societies came into existence to form pressure groups to get stronger action. But generally, and this is always the pattern, legislation lagged behind the reality and, despite the pressure groups, progress was slow: it required the tragic consequences of the Great London Smog of 1952 really to move the government of the day and bring legislation into line with the reality. A year later the Beaver Committee was appointed and its reports paved the way to the Clean Air Act of 1956; this was a major landmark in air pollution control.

This Act is administered by local authorities and deals primarily with reduction and control of emissions of smoke, grit and dust from fuel-burning equipment and sets limits on the density of smoke from chimneys. However, the major innovation contained in this Act was the inclusion of direct controls on smoke from domestic fireplaces. Under the Act a local authority may, with Ministry approval, declare the whole or any part of the district under its control a smoke control area. In 1968 another Clean Air Act modified some of the provisions of the 1956 Act and extended others—it also extended the control under the Alkali Acts.

These Acts have done for industrial effluents what the Clean Air Acts have done for smoke control, and they too have their origins in the middle of the last century when smoke was not the only problem. When the process of making alkali (sodium carbonate) from common salt was first introduced, it produced large volumes of hydrochloric acid and these were discharged into the atmosphere. This caused a public outcry and a Royal Commission was set up to look into it. This Commission's findings led the way to the first Alkali Etc. Works Regulation Act of 1863. It was subsequently extended and consolidated in 1906. Since that time the controls have been widened to cover many more kinds of emission. The Alkali Orders now include nearly sixty kinds of industrial works and cover nearly forty different noxious or offensive gaseous effluents. Under these Acts, the defined processes, which cover mainly the electricity generation, cement, ceramics, petroleum and petrochemicals, other chemicals and iron and steel industries, must not be operated until the owner has obtained a certificate of registration, and this must be renewed annually. After registration, the standards laid down must be maintained continuously. The Government is shortly expected to institute orders to bring under the Alkali Act additional processes and to tighten up the standards still further.

The statutory controls over air pollution are vested in the Clean Air Councils of England, Wales and Scotland. The success of the Clean Air Acts cannot be doubted. The figures testify to this claim: in 1938 smoke emissions amounted to some 2·75 million metric tonnes; thirty years later they had been reduced to 0·84 million metric tonnes—resulting in a visible improvement in Britain's cities.

Pressure on national and local authorities to tackle air pollution must be kept up. New methods of combating it must be found; this is expensive, but in view of the facts (Chapter 10) we have to find the money. Research into air pollution problems is currently being carried out at a number of laboratories belonging to the Government and to industry, and in the universities.

In Chapter 10 we saw that emissions from vehicle exhausts have become the major factor in air pollution. Consequently there has developed a world-wide opinion that these exhausts must be strictly controlled. In the United Kingdom we have as yet no really effective legislation to deal with this problem. There are no limitations on the amount of carbon monoxide that may be emitted, and in a White Paper published in May 1970 it was claimed that there was no evidence that carbon monoxide in the streets has any adverse effects on health or environment.

Other countries are more concerned. The European Economic Commission, for instance, has prepared a standard, which its members may adopt, requiring a 10 per cent reduction in carbon monoxide emissions, and from the autumn of 1971 the maximum level of carbon monoxide permitted in idling exhaust gases from new cars is 4·5 per cent. The Japanese Government have also approved this level for vehicles in their country. However, the nation that has really gone to town on the regulation of vehicle emissions is the United States, where it is recognized that these now form the greatest single group of pollutants.

In the U.S.A. efforts to curb air pollution also go back to the nineteenth century. In 1881 the cities of Chicago and Cincinnati passed smoke control laws; other cities followed suit and by 1912 twenty-three of the twenty-eight cities with populations of 200,000 or more had passed similar laws. However, little overall impression was made. When smoke pollution reached its height in the 1930s, '40s and '50s, more legislation was brought in; this, combined with a gradual reduction in the use of coal as a fuel, did make some impression, but there was little recognition of the broader aspects of air pollution. The trouble lay in the general apathy towards the problem; people tended to equate air pollution with smoke, and the, by then, notorious Los Angeles smog was considered peculiar to that city and not as an indicator of what could happen elsewhere. The first Federal legislation was introduced in 1955—this provided an annual 5 million dollars to the Public Health Service department of the Department of Health, Education and Welfare for research, data collection and technical assistance to State and local governments.

However, demands for improving air quality did increase and eventually led to the Clean Air Act of 1963, which provided grants to air pollution agencies for control programmes and also provided Federal authority to attack inter-State air pollution problems. The Act was amended in 1965 to permit, for the first time, national regulations on emissions from motor vehicles. Standards were set, but despite them carbon monoxide levels will decrease only to about 1985, for after that time the sheer number of cars will cause the levels to increase again. If the standards become even tighter, then society may have to resort to vehicles powered by other means, for example steam or electricity. In the meantime someone—the user—has got to pay for the control devices on petrol-driven vehicles. Concomitant with these regulations on exhaust emissions, limitations on the amount of lead allowed to be added to petrol will have to be set. In the United States it is planned to reduce lead levels gradually to zero by 1974. Petrol containing lead will still be required for the older high-compression-engined cars still running, but a tax will be imposed on the lead so that the new fuels are competitive.

In the matter of exhaust pollution control, California has been constantly a step ahead of the Federal Government. To conform with California State laws, 1966 vehicles were required to be fitted with control devices, two years before this was required at a national level. And in 1970 a Bill was introduced to ban the sale of internal-combustion-engined cars completely by 1975. This Bill went through the Senate but was killed by the Legislature in Committee.

The United States furthered its attempt to control air pollution by the introduction of its Air Quality Act in 1967; this provides the basis for the efforts now being made to clean up air pollution in the 1970s. In 1970 the President put forward more proposals for improving the air quality, among them proposals to apply the air quality standards throughout the country, legislation to enable the Federal Government to enforce them if a State failed to meet these standards, and proposals to apply standards for harmful pollutants such as asbestos, cadmium and beryllium.

Already standards have been issued for sulphur oxides, particulates, carbon monoxide, hydrocarbons and photochemical oxidants. Standards of lead, nitrogen oxides, fluorides and polynuclear organic compounds were planned for 1971, and the National Air Pollution Control Administration is at present studying thirty different pollutants to examine their effect on health. NAPCA carries out extensive research, development and demonstration programmes on all aspects of air pollution.

It also operates its own air monitoring system, which consists of over a thousand air-sampling stations dotted over the country, and it also assists the two thousand State and local systems. But very few of the stations monitor continuously, and the distribution of the stations throughout the country is such that it is very difficult to determine trends in air quality, and the trends that are observed may be misleading because they may not cover all pollutants. Much more monitoring is required. Despite all the legislation there is room for improvement here as well. So much more could be done, and in the words of the First Annual Report of the Council on Environmental Quality, 'The benefits which can be derived from greater control of air pollution far outweigh the costs of the control measures'.

A number of countries now have air pollution control programmes, and on the broader scene several international organizations are studying this problem, which is no respecter of boundaries. We have already mentioned the EEC standards for car exhausts. The World Health Organisation in a comparative study of air pollution over London and Washington is looking particularly at sulphur dioxide and dust particles. It is co-operating with UNESCO in the International Atomic Energy Agency's world-wide sampling network for checking radioactivity in precipitation and its work on techniques for measuring atmospheric and other environmental radioactivity. The World Meteorological Organisation has launched a World Weather Watch to improve the collection of meteorological data for forecasting and for studies of atmospheric pollutants, and the International Biological Programme is also looking into ways of identifying pollutants in the atmosphere. The United States, Germany and Turkey under the auspices of NATO's committee on 'The Challenge of Modern Society' are collaborating in a comparative study of air pollution in two cities: Frankfurt, as an example of a large industrial city, and Ankara, where there is a wide use of a coal with a high sulphur content.

Next to air pollution in importance is the pollution of our rivers and lakes, and as would be expected efforts have been made from time to time to prevent or control it. An attempt was made in the fifteenth century to stop the use of the River Thames as a dustbin, but it met with little success. The rivers were too convenient as dumping grounds and this consideration far outweighed any recognition of the damage this was causing, and the population living on the river banks had to learn to live with the filth and the smell. But as these populations increased with the arrival of the Industrial Revolution the problem, like that of air pollution, could not be ignored any longer. By the middle of the 1800s British rivers had deteriorated so much that the health risk reached an all-time high, and outbreaks of cholera prompted an investigation. In 1866 a River Pollution Prevention Commission reported on the condition of the rivers Thames, Lea, Aire and Calder, and further Commissions between 1868 and 1874 studied rivers in other industrial areas. Their reports are a landmark in pollution control. The 1874 report in particular defined for the first time the permitted polluting limits of liquids entering streams, and laid down regulations on the disposal of solid matter dumped in rivers and streams. The work of the Commissions led to the Rivers Pollution Act of 1876, but its provisions were restricted, for it did not define polluting effluent and did not cover pollution of tidal waters except where there was a public health hazard. Prior to this, in the 1860s Acts of Parliament were passed to protect salmon fisheries specifically, and these were subsequently extended in 1923 to cover all fresh-water fisheries, their spawn and food.

A further attack on the problem was made by the setting up of a Royal Commission on Sewage and Sewage Disposal which over the years 1898 to 1915 issued nine reports which were to become the accepted standard on the subject. Among the Commission's recommendations were that standards of purity should be applied to effluents in respect of suspended solids and the oxygen-absorbing power of the effluents, and also that water resources should be the concern of local Boards. These reports were endorsed by the Joint Advisory Committee on River Pollution (1928–37), but unfortunately there was no legal enforcement. The Advisory Committee on River Pollution was reconstituted by the 1945 Water Act and one of its first tasks was to set up a sub-committee to look specifically into pollution matters. The result of this was the Rivers (Prevention of Pollution) Act of 1951 which marked a significant step forward, incorporating as it does the recommendation of the early Royal Commission, albeit over fifty years later. This Act empowered River Boards, '(a) to prescribe standards for effluents, (b) to ensure that new openings for effluent discharge could only be carried out with their consent, and (c) to enforce any provisions of the Act in tidal waters by order of the Minister of Housing and Local Government.' Enforcement of the Act was left to the individual River Boards as the same standards could not be expected in a river in an industrial area as in a rural fishing river. In 1960 the Clean Rivers (Estuaries and Tidal Waters) Act extended legislation, as the title suggests, to estuaries and tidal waters but it only covered discharges beginning *after* the time it became law. Many authorities which had been dumping wastes into these waters for years previously could go on doing so. As yet nothing has been done to remedy this, and it is a gap in the legislation which must be filled.

Similarly, industrialists are not compelled to disclose what is in their effluents, and many pollution officers believe that too many manufacturers are sheltering under the legal umbrella which is offered by a section of the Rivers (Prevention of Pollution) Act of 1961 which prohibits River Boards from revealing either the content or the standard of a company's effluent. And as we saw in Chapter 7, with the increasingly complicated effluents that are being discharged and the very real possibility of interaction between effluents, legislation to revoke this protection will have to come. There are fines for the deliberate discharge of untreated effluents to water courses, but these are notoriously low, and the law does not cover accidental discharges. There is a strong movement to include such discharges in the law because many culprits avoid the existing penalties by saying that the particular discharge concerned was an accident.

There is no doubt that the standards of river quality are improving and in some cases there have been dramatic improvements, in the Thames for example; so much so that within the next five years fish might well be seen again in the middle reaches.

River authorities are currently conducting a major study to determine what improvements have occurred since 1958 and to estimate how much it would cost to remove the pollution that exists. A particularly comprehensive study is being made of the River Trent, a badly polluted river, in a combined effort by the Trent River Authority, Water Research Association, Water Resources Board, Water Pollution Research Laboratory and a number of government departments. The study is aimed at making the best and most economical use of Trent water. The Local Government Operations Research Unit at Reading has devised the most complex mathematical river model ever constructed, based on the Trent studies, and this is to be put through practical computer trials to establish a basis for an integrated water policy. The model is designed to give a dynamic representation of water quality throughout the river and takes account of the different pollutants, variants such as high and low flow rates due to seasonal changes, variations in temperature and variations in the natural purifying capacity of the river. It can also predict the effect on the river of a particular user and it will be possible to calculate costs likely to devolve on existing users in order to maintain river quality resulting from a new user up-river.

Much of Britain's research into water pollution problems is

carried out in the government Water Pollution Research Laboratory at Stevenage. Currently more than half of its effort is devoted to improving methods of forecasting the effects of pollution on the quality of natural waters and on their plant and animal life. This work should provide a more effective means for planning the development of water resources and of controlling pollution in them. The laboratory is also working on the optimum siting for outfalls for sewage from coastal towns, the prediction of the effects of effluent discharges, the influence of the enrichment of natural waters with inorganic forms of nitrogen and phosphorus on troublesome growths of algae, the treatment of effluent from a variety of industrial processes, and also sludge treatment processes for both industrial effluents and sewage.

The current cost of river purification in Great Britain is about £100 million a year, but despite all that is being done, this is only a part of what should be done. A spending rate of £300 million would be far more realistic. As we saw in Chapter 7, three out of five of Britain's sewage plants are producing effluent below standard, despite an increase of 40 per cent in capital expenditure on sewage systems since 1964. Unfortunately in 1968 local authorities were advised by the Government because of economic difficulties to limit their expenditure to schemes needed for urgent reasons of health or for new industrial or housing development. They could not, however, take any action that might worsen the condition of a river. The Government have now withdrawn this advice and given the go-ahead to allow both the river and local authorities once again to undertake positive purification measures.

In 1969 the Government set up a special Working Party to examine the public health, amenity and economic aspects of the various methods of sewage disposal. The report of the Working Party was published in 1970 and showed up many of the deficiencies. The report stresses the need for sewage disposal to be regarded as a vital industry rather than a convenient means of forgetting about unpleasant wastes. Among its recommendations are that (1) boats should not be allowed to discharge sewage into inland waters used for recreation; (2) planning permission for new housing should be withheld until adequate sewage facilities are available; (3) trade effluent discharged into public sewers should be subject to a charge if it contains toxic substances that make treatment more expensive; (4) surface water drains should be separated from foul sewers; and (5) crude sewage should be discharged into the sea only through diffusers or long outfalls and after adequate screening.

The main point made by the Working Party, however, is that if we are to have sufficient good quality water to meet the demands of a rising population and of industry, there must be a change in the administrative structure. At present, water management is the concern of a large number of different authorities: in England and Wales alone there are 79 municipal water undertakings, 106 water boards, 30 river authorities and 21 joint sewage boards. The Working Party says that while much could be done by numerous small improvements and by closing legislative loopholes, the stage has been reached when a reduction of the number of authorities must be made, based on a scientific understanding of the complete water cycle. It suggests that there should be a central planning organization, or expanded Water Resources Board, the main body, which would also run the Water Pollution Research Laboratory.

The attention paid to water pollution problems and legislation going back many years have put Britain in the forefront of pollution control. It is to be hoped that all the plans and promises are not just pie in the sky but will come to fruition, and that the day will not be too distant when all effluent, domestic or industrial, will be treated and rendered harmless.

In the United States it was not until 1948 that the first water pollution control was introduced, and this was only temporary. Permanent legislation had to wait until 1956, when the Federal Water Pollution Control Act was passed. This authorized planning, technical assistance, grants for State programmes and construction grants for municipal waste treatment facilities. The Act was amended in 1961 to extend its enforcement powers and to provide increased construction grants. It was further amended in 1965 to establish the Federal Water Pollution Control Administration and this amendment provided for the establishment of water quality standards and put forward plans for the clean-up of all inter-State and coastal waters. Each State was required to establish standards for their inter-State waters, which would then be approved as Federal standards by the Secretary of the Interior. If they failed to do so, the Secretary of the Interior was given powers to establish Federal standards and to enforce them. The difficulty in implementing this Act has been that not all States have set the same standards, so the aim of getting comprehensive water quality standards throughout the U.S.A. has not yet been reached. Progress is being made, but there is a need to control more strictly pesticides and industrial effluents. Early in 1970 the President sought authority to require the States, with Federal approval, to set specific effluent discharge requirements as part of water quality standards. This could provide a standard which could be enforced. Several States, particularly Illinois, Pennsylvania and California, are making a very determined effort to prosecute polluters, and the Federal Government has also brought some successful actions. Further legislation was introduced in 1966 in the Clean Water Restoration Act which provided more Federal money for treatment plants.

Since 1957 the Government has put $1,500 million and individual States and cities $6,400 million into building new and improving existing municipal sewage treatment plants. But with the increasing population and the increasing amount of waste produced per head, many plants are completely inadequate or have become obsolete despite this expenditure. Needs have far outpaced the finance available. It was only in 1970 that the Federal contribution made a significant step forward when $800 million were granted as against $214 million in 1969.

If water standards are to be enforced there must be some method of water quality surveillance. The Federal Water Quality Administration (established out of the Federal Water

Pollution Control Administration by the Water Quality Improvement Act of 1970, which provided even tighter controls over oil pollution, vessel pollution and pollution from government activities, as well as widening the scope of the earlier legislation), the United States Geological Survey and State Pollution Control Agencies have been developing a national monitoring system. At present this covers 20 per cent of the fresh-water supplies, but ultimately it will grow to include thousands of stations and carry out in addition intensive attacks on specific problems. Several other government agencies are monitoring programmes to supplement the data from the pollution agencies.

One of the most serious pollution problems, as we have seen, concerns the Great Lakes and here a large-scale attack is being made on a national, state and local basis involving research, water quality standards control and close co-operation. All States bordering the Lakes have set high quality standards. The four States bordering Lake Michigan have put forward a very comprehensive anti-pollution and clean-up programme for the Lake which, if it is carried out, should be very effective. Canada and the United States are to make 1972 the International Field Year for the Great Lakes, when they will collaborate in a detailed data-collecting exercise of conditions in Lake Ontario. This research will be part of the contribution of both countries to the International Hydrological Decade which began in 1965. Research into water pollution problems is an essential part of control and the American programme is increasing. The Government allocated $61 million in 1971 for research and development of new technology.

Many other countries are intensifying their attack against water pollution. The severely polluted Rhine poses an international problem and to improve conditions an International Commission for the Protection of the Rhine has been set up. Recently the Dutch and the Germans have held discussions to increase their co-operation and plans have been laid for joint monitoring and early warning systems. The Dutch, ever inventive, have devised a very ingenious method for reducing the pollution from river water feeding The Hague water supply. Rhine water has to be cleared of much salt and organic matter before it can be pumped through the pipeline to the city. To make this water potable it is fed into nearby sand-dunes from which the city's main water supply has been drawn since 1874. These dunes provide a natural filter bed and the water quality is much improved. At the same time the extra water replenishes the underground reserves and can be called upon when there are water shortages or if the Rhine water is so polluted that it just cannot be pumped through the pipelines.

West Germany has recently launched a wide-ranging programme against environmental pollution which includes anti-water-pollution measures. Germany is currently suffering from an acute shortage of sewers and effluent treatment plants and only 40 per cent of the country's domestic and industrial effluents are being treated adequately. The Government is already hastening the construction of 5,000 new treatment plants costing some 600 million DM, but this is only a start. Within the next thirty years water requirements are expected to double and it will be necessary to update the anti-pollution regulations and to tighten controls. Although local authorities will be given loans, grants and tax concessions, water rates will have to rise to pay for the necessary treatment schemes.

On an international scale, UNESCO, WHO, UN, FAO, International Atomic Energy Agency, OECD, Council of Europe, COMECON, are all tackling water pollution problems and a European Federation for the Protection of Water has been set up.

While nobody doubts that it is important to clean the air and water, the same measure of agreement is not found when we come to examine the land. There are many who feel that the pesticide threat has been exaggerated and even more who doubt that the problems of derelict land and waste disposal are major issues. We cannot get away from the fact that the introduction of pest controls has brought enormous benefits in terms of food production, but it is equally a fact that the increasing use of chemical pesticides has brought with it one of the major threats to wild life and Man.

In Britain official pesticide control dates back to 1954 with the setting up of the Advisory Committee on Poisonous Substances Used in Agriculture and Food Storage. Its terms of reference were widened in 1964 and the name was changed to Advisory Committee on Pesticides and Other Toxic Chemicals. During 1964 the committee carried out a review of persistent organochlorine pesticides and recommended that voluntary restrictions on the use of aldrin, dieldrin and heptachlor should be introduced in certain cases and that the remaining approved uses of these particular chemicals and of the others of the organochlorine family should be reviewed again in 1967. A further detailed survey was launched in 1967 and the findings published at the end of 1969. Although in its report the committee recognized that some of these persistent chemicals were now an integral part of the environment, they had no evidence that they were in fact directly harmful to man and thus they did not consider that a complete withdrawal of their use was necessary. However, they did comment that the presence in the environment of these persistent chemicals was undesirable and that an attempt should be made to curtail their use. The British Government is, in fact, expected to bring in tighter regulations.

Some other countries have introduced even stricter controls. The United States has recognized the dangers for a number of years and in 1969, when the dangers of indiscriminate pesticide use really became apparent, control regulations were tightened up considerably. In July the use by the Department of Agriculture of nine pesticides was suspended pending a month-long review of their contamination, and the use of DDT, dieldrin and heptachlor was severely reduced or completely banned. In November the then Environmental Quality Council set up a sub-committee on pesticides and the Secretary of Agriculture announced a complete ban on all use of DDT for house and garden pests, shade-tree pests, pests in aquatic areas and tobacco pests. Subsequently in January 1970 an inter-departmental agreement covering agriculture, interior and health, education and welfare was signed which provided for the strengthening of the review of pesticide regulations in relation to the protection of human health and the environment. Further regulations followed: the suspension by the Department of Agriculture of forty-two alkyl-mercury fungicides for seed treatment, the banning of aldrin and dieldrin insecticides in aquatic environments by the Department of Agriculture, the suspension of the use of liquid forms of 2, 4, 5-T for home use and all forms of it for use on lakes, ponds and ditches by the Departments of Agriculture, Interior and Health, Education and Welfare and the complete suspension of DDT, aldrin, 2, 4, 5-T, dieldrin, endrin, DDD, mercury compounds and nine other pesticides on public lands controlled by the Department of the Interior. In addition, individual States have their own regulations and many have been considering tightening up their restrictions. Michigan, for example, has banned the use of DDT and Arizona has restricted its sale.

The story has been repeated in other countries. Canada has recently banned DDT in more than 90 per cent of the operations in which it was used. It was banned from non-agricultural uses and permitted for only twelve particular farm products, and from the beginning of January 1970 it could be used only under emergency conditions in forests and public parks. The tolerated DDT level in foods was also sharply reduced. Sweden has declared a temporary ban on DDT in order to discover whether the DDT residues found in that country have come from internal or external sources. This also applies to aldrin and dieldrin.

The pesticide problem cannot be regarded solely as the experience of individual countries: it is truly an international concern—the DDT residues in the fat of Antarctic penguins are sufficient evidence for this. Pesticides are no respecters of national boundaries. The problem has been studied by the World Health Organisation, Food and Agriculture Organisation,

Organisation for Economic Co-operation and Development (OECD), Council of Europe and the International Union for the Conservation of Nature. Chemists and biologists working under the auspices of the OECD have been studying the reliability of methods of analysing pesticide residues and the feasibility of an international monitoring scheme. As political and economic difficulties can arise if different nations have different safety regulations for the use of pesticides, the need for standard registration schemes is recognized, particularly between nations at the same level of development. The Council of Europe has made a particular effort in this direction in its booklet *Agricultural Pesticides*, prepared by the Working Party on Poisonous Substances in Agriculture, first published in 1962 and revised in 1969.

The withdrawal of dangerous pesticides means that others, less dangerous, must be found to fill their place. There *are* degradable chemical pesticides, but there is increasing research into non-chemical pest control such as the use of predators and parasites or the use of pathogens—viruses, bacteria and fungi. There are, however, dangers here. The introduction of new species to a habitat can be risky: for example, the mongoose which was introduced into Jamaica and Puerto Rico to control rats has now itself become a pest, and pathogens may be harmful to human health. One of the safest ways of non-chemical pest control so far is by the selective breeding of resistant species of plants; much more success has been achieved with this method than with the others. It is worth noting that about 80 per cent of all pesticides are used to control fewer than a hundred species of pest organisms and it seems, therefore, that if non-chemical methods can be found, the pesticide load could be considerably reduced.

It is only comparatively recently that efforts have been made to try to reclaim derelict land in this country, and as we saw in Chapter 6, government help for this has been available only since 1967. Successful tree planting on and seeding of spoil-heaps have improved the landscape in several areas and more and more schemes are in hand, but so much more could be done. The successful ventures can serve as an example and a reproach to less enlightened areas.

Although dereliction is the most obvious eyesore, and the legacy of exploitation of resources without heed to the consequences is the root cause, much of our remaining countryside would suffer a similar fate if it were not for a number of admirable organizations which act as watchdogs on what we do to it. Some are government or government-supported organizations, others purely voluntary, and are aimed at conservation of what we have. In 1949 the Government set up the National Parks Commission, now the Countryside Commission, which administers the ten National Parks covering 9 per cent of England and Wales. Scotland has its own Countryside Commission established in 1968. Also in 1949 the Nature Conservancy was set up to advise on nature conservation, establish and run nature reserves and to carry out research. Altogether there are 128 nature reserves covering a quarter of a million acres under its control. The Forestry Commission was set up after the First World War primarily to boost timber production, but it is now becoming more ecologically conscious and doing much to establish new stands of timber. Of the voluntary organizations the Civic Trust, founded in 1957, encourages high quality in architecture and planning, preserves buildings of historic interest and aims to protect the beauty of the countryside; the Council for Nature, the national representative body of the natural history movement founded in 1958, was set up to promote the study and conservation of nature; the Council for the Protection of Rural England, founded in 1926, aims to protect the countryside and to rouse public opinion to awareness of the dangers and also acts as an advice and information centre; the National Trust, founded in 1895, has been, and is, extremely successful in preserving places of historic interest or outstanding natural beauty for the benefit of the whole country. The National Trust for Scotland, founded in 1931, has a special committee on Wildlife Preservation. The 'Countryside in 1970' movement, originally started in 1963, was a series of conferences convened by the Duke of Edinburgh bringing together representatives of the 2,000 organizations to consider changes in the countryside. The conferences at least have had the effect of increasing public awareness of their environment in the widest possible sense. At its culminating conference in 1970 the movement set up a Committee for Environmental Conservation (CoEnCo) to continue to co-ordinate the work of the independent bodies.

The United States has been concerned with conservation since just before the First World War, but until recently there has been relatively little research to back it up. The country has organizations similar to those in Britain—Forest Service, National Parks Service, and Fish and Wildlife Service, and a National Wilderness Preservation System, plus numerous voluntary organizations. In 1969, $142 million were spent by Federal and private organizations on wildlife research, management and habitat protection.

Protection of the environment would be eased considerably if we could solve the problems of waste disposal, but with the present population and the concomitant production of wastes, this is a very formidable task. Simple dumping is just not acceptable and yet it is a policy that still exists, even though the gross results of ugliness, pollution and health hazards are only too obvious. Fortunately awareness of the importance of waste disposal is growing, and those responsible for it are beginning to realize that it is no longer a bottom-of-the-list priority but a vital necessity. The over-exploitation of resources is beginning to take its toll and virgin raw materials are becoming in short supply; this has brought a realization of the need—which will become more and more important—to recycle these materials from waste. In Britain some enlightened councils are in fact making a determined effort to salvage their refuse. In a House of Lords debate in 1970, Baroness Emmet of Amberley quoted the example of Worthing Corporation which, she told the House, salvages waste paper, tins, ferrous and non-ferrous metals, bones, fat, bottles and rags. The paper is packaged, the metal compressed, the rags sorted, and compost is made from sewage sludge. What is left is used as covering material for a dumping tip and this is landscaped. Nothing is wasted and the Corporation estimate that over the past five years they have saved imports worth £250,000. Much of the success has been due to the efforts of the chief engineer, who carried out a personal publicity campaign in training householders in how to dispose of their refuse. The Worthing system even uses waste to power the disposal plant—methane gas is extracted from the sewage to provide the necessary electric power.

An increasing number of systems utilize the gases produced in incineration to power the incinerators and to produce electricity for other uses. London's £10 million refuse disposal plant at Edmonton, which gobbles up 1,800 tons of rubbish every day, produces enough electricity for 25,000 households. The refuse is reduced to ash and metal, the latter being extracted with magnets. The plant will earn some £250,000 a year with its by-products, and there is no air pollution. Any grit is separated out and the only effluent is steam. The American CPU-400 system works along similar lines; this is a fluidized bed incinerator which burns solid waste at high pressures. The hot gases produced power a turbine which drives an electric generator. This system can cope with 400 tons of refuse and produces some 15,000 kW of power daily. Nearly three-quarters of the heat available in the turbine cycle can also be used for steam raising and sewage sludge drying.

While the bulk of responsibility for waste disposal has rested with industry and local authorities, the Government is increasing its interest and introducing legislation to improve standards. In Britain, the Government has set up special Working Parties to look into the disposal of toxic wastes and of household refuse in particular. These follow an earlier Working Party set up in 1967 which considered refuse storage and collection. Unfortunately this did not have the impact it should have, local authorities using the excuse of shortage of money to delay implementing its recommendations and the Government not applying the necessary pressure to get authorities to abandon outdated methods. However, the 1970 White Paper, 'The Protection of the Environment', promises that following the publication of the findings of the current Working Parties, the Government will introduce 'such improvements in the law as they consider to be desirable'.

In the matter of the recycling of wastes the Warren Spring Laboratory has been carrying out research into the recovery of a number of materials from alloy scrap, chemical effluents and sludges. The main idea of the scrap work is to adapt processes already used in the primary metal-producing and extraction industries so that they can be used in the secondary metal industry, and it is also hoped to encourage the treatment of scrap and other residues which are not being treated at present in this country. Research workers in industry and the universities are also turning their attention to problems of waste recycling and the production of new materials from waste. Plants are already planned and operating in the U.S.S.R., Poland and Czechoslovakia for the production of glass-ceramic materials from blast furnace slags. These materials find particular application in the construction industry. Britain's equivalent is not yet on the market, but this is mainly due to the slowness of the building industry to adopt new materials, and to the regulations imposed on the industry by local authorities, building societies and insurance companies.

In Chapter 6 I mentioned the special disposal problems posed by plastics which do not degrade naturally. This is a field crying out for research and efforts are being made in this direction. Scientists at Birmingham University have claimed that they have found a way of getting polythene and polypropylene plastics to degrade after being thrown away—a sort of delayed shock if you like, but the work is still in the very early stages.

In America, as in the United Kingdom, most of the primary responsibility for solid waste collection, processing and disposal falls on the local authorities, but in 1965 the Federal Government took a hand. The Solid Waste Disposal Act gave the Federal Government responsibility for research, training, technical assistance and demonstrations of new technology, and provided grants for State and inter-State solid-waste planning programmes. Emphasis is placed on conserving natural resources by reducing waste and unsalvageable materials, and on solid waste economy. Under this Act the Bureau of Mines was given increased authority to look into problems of mineral, metal and fossil-fuel wastes. The Bureau of Solid Waste Management is charged with administering the Federal programme for solid wastes from all other sources.

In the United States it is being realized that waste disposal poses problems increasingly outside the control of the immediate local authority and that a more regional control is necessary. This has its own problems in the difficulty of getting co-operation between authorities, which tend to guard their own interests and position jealously. Direction can only come from above.

The same applies to the marine environment. At the present time each country lays down its own regulations, if any, regarding the pollution of its territorial waters. But as in the case of atmospheric pollution, marine pollution is an international problem and legislation is needed urgently to control it. In international waters the Inter-Governmental Maritime Consultative Organization (IMCO) operates particularly for the control of oil pollution. IMCO conventions are bringing under increasing control the operations of tankers and other ships at sea and the liability of the owners to pay for damage caused by oil spills. The 1954 International Convention for the Prevention of the Pollution of the Sea by Oil lays down conditions under which oil may be discharged at sea, and puts limitations on discharges within certain prohibited areas, such as the North Sea, the English Channel and much of the North Atlantic. The 1969 Amendments to this Convention made all the seas a prohibited area and limited discharges to an amount which had been shown experimentally not to produce persistent oil slicks. Most oily residues will have to be kept on board ships or discharged to shore installations. The Convention also provides for a system of records and inspection to assist enforcement. The British Government has ratified these amendments and has presented to Parliament the Oil in Navigable Waters Bill which will provide the legal backing. Under the Bill the present maximum fine of £1,000 is to be increased to £5,000 on summary conviction of a master whose ship illegally discharges oil into the sea. Two 1969 Conventions make it easier for governments to intervene in order to protect their coasts in case of accidents and tighten up the insurance cover which tanker owners are obliged to take out. Owners are now liable for a much greater proportion of the damage caused by an oil spill than they were.

Many research organizations, including the Water Pollution Research Laboratory and the oil companies themselves, are investigating ways and means of destroying oil slicks when and where they occur. Both the British Petroleum Company and ICI have produced chemical dispersants which are less toxic than those used in the case of the *Torrey Canyon*. Investigations into methods of making the oil sink are also being carried out in many maritime countries, but particularly Britain and France. In the latter country Technocéan has designed a ship which could skim an oil slick off the sea, separate it from the water and carry the oil either to a tanker or to a land installation. Such ships are too expensive for one country to operate but the cost could be shared by a number of countries, for instance if one was stationed in the Channel, the expense could be shared by Britain, Belgium and France. This way skimmer ships would be available wherever oil slicks occurred.

The disposal of radioactive wastes is one of the major problems of modern society and this has yet to be resolved. In the meantime we have to put up with the idiotic policy of dumping in the marine environment. We have to accept that radioactive waste must be dumped, but it would be safer to dump it in deep mine shafts which could be sealed. And this policy of dumping is not confined to radioactive wastes. It extends to anything;

even while making noises about a cleaner sea, particularly the North Sea, the British Government authorizes dumpings of industrial wastes there. There is an urgent need for an international regime of the sea, but its likelihood is inversely proportional to its urgency.

Most of the environmental problems are international, and although agreement has been reached in principle in many areas, there are wide gaps between the signing of the agreements and putting them into practice. It is only in the last ten to twenty years that some of the problems have been recognized as such, and even now there is far too little action. The political and economic considerations are still the main criteria for governmental action. Ecological principles are relegated to a minor position, although it is currently fashionable for governments in the Western countries to pay some attention to environmental problems. In Britain the Labour Government in 1969 set up a Royal Commission to look into environmental problems in the British Isles. The White Paper of 1970 said nothing that was not already known. The First Annual Report of the United States Council on Environmental Quality published in August 1970 is an excellent piece of writing and will, I am sure, be used as essential reading by students for years to come. However, there is a common theme that runs through both documents and that is the need for centralized control of environmental quality; the onus is on governments. It is clear that up till now the thinking, where there has been any, on environmental problems has been fragmentary, and consequent attempts at solving them, where they have been made, have been piecemeal. In most cases the initiative to do something about the problems has come from individuals or pressure groups, such as the voluntary organizations. In the United Kingdom the setting up of the Nature Conservancy in 1949 was a major landmark in the conservation scene. Its scientists have played a most important part in elucidating ecological principles, and more recently their work on pollution and pesticides in particular has been of the highest quality. Sites of Special Scientific

Interest (SSSI) may to most of us be rather academic, but the 'building blocks' of the ecological structure are discovered in such places.

In this chapter I have not referred to the particular problems of the city because they are so numerous that their recording would take many volumes. Perhaps the first requirement in the understanding of the city as a habitat is for the ecologists to move in. For if we are to understand Man in a habitat, we must study his interaction with his habitat, and today for many millions that is the city and urban sprawl. In the meantime a great deal could be done to clean up the cities if the existing laws on litter were enforced and the penalties for infringement are made to have some bite. Regarding personal pollution we can only hope that such things as the anti-smoking campaigns will have increasing success, and that governments will be stronger in dealing with drugs and 'drug pushers'. Perhaps there is a case here for capital punishment, for these people who play on the weakness of others and live by degrading and ultimately destroying their fellow human beings are not worthy of our consideration.

We have seen some of the things that have been done to ameliorate the pollution of our environment, we have seen how governments were forced, either by events or pressure groups, to take action. More recently, as the problem has again grown acute and pressure from the more alert citizens has made governments look more closely into environmental matters, an increased effort has been made. Unfortunately governments have many ways of taking the steam out of the situation, and the public must be on its guard against these stratagems. However, we must acknowledge the improvements that have been and are being made. For 1972 the United Nations has convened a Conference on the Human Environment to be held in Stockholm. This could be an important step in international co-operation, but experience teaches us not to expect too much from it.

A lot of work has been done, and obviously a great deal more will be done, but unless it is carried out on a larger scale and in a co-ordinated way the problem of pollution will not get better, it will get worse. The production of waste far outstrips its elimination and the drain on the Earth's natural resources will inevitably go on while the world pursues an economic policy of growth. Legislation will always lag behind the need, until new attitudes are forthcoming, and this will only be when the majority of people become aware of the threat to their welfare. All efforts are ultimately frustrated by the attitudes of those whose interests lie only in the continuance of the present open-ended economic system and hold on to the attitude that all that is needed is money: money is not enough to fight the wrongs—prevention is better than cure—often there is not a cure! Yes, Mercutio, 'twill serve but 'twon't do!

CHAPTER 16

Ultimatum for survival

Few thinking men doubt the precariousness of the situation in which Man finds himself today. All around are signs enough that things are going wrong on our native planet. In this book we have examined the effect Man has had on the natural environment in which he himself evolved. We have seen how during the last two hundred years especially his material success has led not only to the degradation of the environment but also to a lowering of the quality of his own life. We have seen how the triumphs of modern medicine and surgery have led to perhaps the greatest dilemma that has ever faced the human species, whether to let die for the sake of the species, or whether to save for the individual. The world is divided politically into roughly two ideologies; it is also divided economically between the 'haves' and the 'have-nots'. The ideologies absorb enormous quantities of materials and men in defending political illusions. The wealthier nations are getting wealthier and the poorer nations poorer. The genius of modern Man is for making mistakes!

The errors of the past are everywhere being perpetuated as the developing nations try desperately to copy the established industrial states, and the latter offer the young nations the 'benefits' of their way of life. The 'Establishment', both religious and secular, is collapsing because its foundations are being swept away in a current of change. This is not entirely the fault of the politicians, the religious leaders or the bureaucrats; mankind has never faced such a situation before. It is

completely new and brings with it new values. Its complexity
is overwhelming.

We have time, but only a little, in which to pull ourselves
together and to decide on survival or extinction. The Earth's
limits have been defined. We know where we are and what we
have got, and we know that there are too many of us. Within
the next decade the greatest decisions that men have ever had
to make must be made. The first will be how to control their
reproduction. In the backs of our minds we know that something
must be done but we secretly wish that the decision would be
taken out of our hands. The family has to be planned, births
have to be reduced. Faced with the appalling population problem, the Papal Encyclical, *Humanae Vitae*, was one of the cruel-
lest blows aimed at the survival of mankind. Countless millions
were doomed by its proclamation to misery and starvation. We
can no longer indulge in outmoded dogma.

While we decide what to do about our numbers, we must
put our 'spaceship' in order. The chaos must be brought under
the control of a disciplined crew. The first thing we have to
assess is the present state of the life-support capability of the
planet. Everything we need to support its crew and passengers
has to be found within its confines. It was originally 'designed'
to support a balanced ecosystem which included men. Now
there are too many passengers!

To survive we need air, water, food, fuel and minerals. We
must know how much of these we have, how long they will last
at the present rate of consumption. We have to suppose that
mankind will live on this planet for thousands of years and we
can presume that his technological ability will increase so that
in some far distant time he will be able to perform the miracles
that we can only dream about. It is our responsibility to ensure
that Man survives his present difficulties. We must ensure that
the human population is geared to the planet's resources. When
we have assessed these, we shall be able to say how many
of us the planet will support and for how long. The work of
compiling this knowledge has begun but it needs to be stepped

up. Concurrent with this must go the work of cleaning up the environment. This will be a difficult task for it will take not only all the ingenuity we can muster but all the will power too.

Ironically, it is our interest in space that has made us aware of the smallness and uniqueness of our planet while at the same time providing us with the technology for survival. Many have decried the money and resources spent on the space programmes of the United States and the Soviet Union, contending that the money would be better spent on the starving millions. Better, but for how long? Space science has provided us with a unique means of studying our world. Satellites can search for minerals, study the land surface and the seas' movements and provide a continuous record of the changing moods of the atmosphere and information on the intensity and nature of solar and cosmic radiation. With their aid Man is beginning to understand the planet's energy balance and its atmospheric and oceanic 'machines'. Satellites will be improved, their remote sensing devices refined, and eventually they will be used for ecological studies. The trouble with satellites, as I observed in Chapter 11, is that they become obsolete and end up as space junk, but as the techniques are further developed perhaps the technicians will produce recoverable satellites.

Aircraft have for some time played an important role in the exploration of the Earth's land resources, and new techniques will increase their contribution still further. Thousands of flying hours have been logged in geophysical surveys which are important for the discovery of oil, water and mineral resources. Supported by the work of highly trained geologists and surveyors on the ground, aircraft have helped to provide accurate maps of the most inaccessible areas of the world. Aircraft can do in a few weeks what once would have taken years by men working on the ground. The surface minerals of most parts of the world have been worked out, and the search now is for those minerals that lie beneath the surface; for this highly sophisticated techniques have to be used. Radiation detectors reveal the presence of radioactive ores such as uranium and radium;

the magnetic ores of iron can be picked up by magnetometers, and infra-red detectors can show minute differences in surface temperature of the land and oceans which give clues to the type of vegetation and plankton distribution. Aerial photography in combination with carefully thought-out programmes of field study can provide the essential data for rational land use and urban planning.

At sea, research ships use vastly improved techniques for exploration of the ocean and ocean bed. The war-time use of sound waves, or sonar, to detect submarines has been turned to peace-time advantage—with its aid a survey ship can produce a profile of the ocean bed. And where this is not accurate enough, a range of submersibles is available to take Man to the depths of the hydrosphere. Oceanology is a new science, a synthesis of many disciplines, aimed at exploring and exploiting the resources of the seas, but these are our reserves. We must know what our reserves are, but have we the right to draw on them when we are wasting so much? And what of those to follow us? What are we to leave them?

The living resources of the sea are at least renewable. If managed rationally they could provide food and materials probably for ever, but alas, even here, we have a history of over-exploitation. In spite of international regulations, of inter-governmental organizations, we have failed to protect the world's fisheries: the great whale fishery, a comparatively easy one to control, is dying because its raw material is becoming extinct. The only way to overcome a repetition of such disasters is to put the oceans under tighter international control, perhaps under the jurisdiction of the United Nations. The freedom of the seas has been too often interpreted as a licence to destroy. As we saw in Chapter 8, the territorial waters of most nations are already being spoiled, but the individual nations must be prevented from ruining the waters beyond their own limits. Waters outside these limits should come under the new ocean regime of the United Nations who would allow the exploitation of the biological resources only under licence and with strict safeguards. Under this regime the mineral and other resources of the oceans would be considered as reserves, and not for use by this or many generations to follow.

The United Nations should also be given the responsibility for the quality of the Earth's atmosphere. It is essential that an international organization lay down the rules for its use—the rights of Man should include the right to clean air! The advent of the supersonic jet airliners will present many problems, apart from the sonic boom. At the time of writing, there are projected four hundred daily flights of these aircraft cruising at 1,800 miles per hour at 70,000 feet. Apart from the tons of oxygen which these jets will use up, they will inject 150,000 tons of water vapour into the stratosphere every day, increasing the temperature and cloud as they do so. It has to be decided whether this is justifiable in ecological terms. Likewise, space flights which pierce the atmospheric envelope should be investigated for their cumulative effects on the environment. Perhaps the United Nations could ration the frequency at which space vehicles leave and enter the atmosphere; two nations should not have the right to rip our skies apart.

After our assessment of the life-support capability and the emergency reserves, we have to plan how we are to utilize what is available to us. The first thing we have to organize is the recycling of those metals and other materials which we have already at hand. It should be possible to re-use more than 90 per cent of the metals currently flowing through industry, perhaps 80 per cent of paper and the same for cloth. Plastics that can be used over and over again should not be beyond the ingenuity of the chemist. Recycling of materials is an absolute necessity and no economical argument is valid when set against the long-term survival of mankind. For recycling ensures that there will be a constant amount available, and it would also put an end to much of our pollution.

We have to begin to think of food production in terms of energy production. We must understand thoroughly the heat budget of the planet and we must have a detailed knowledge of the energy flow through the biosphere. From this we should then be able to calculate how much energy is available for our use, another way of learning how many of us the Earth can support.

This then is the basis of our planned future as residents on the planet, and it is our only hope if we intend our species to continue. While the politicians continue carping and the technicians are beginning to gear themselves to the job ahead, we have to start to clean up the mess we have got ourselves into, and this is as much the responsibility of the individual as it is that of industry or government. Recyling and the clearance of pollution go hand in hand and can be achieved only if our wastes are available for collection. It means an end to litter in the streets and public places and in the countryside. It means goodbye to the dumping of the old refrigerator, bicycle or motorcar. It means an orderly and tidy way of life. Governments have the biggest job of all. They have to ensure that the clean-up begins and is maintained by enforceable laws if necessary. Reclamation of derelict land must be made compulsory and the spread of urbanization curbed. The poisoning of the land, the fresh water and the atmosphere must be brought under control and the sea must be protected—it must not be used any longer for dumping anything and everything; the example set by the Swedish Government (Chapter 8) should be followed. Oil pollution of the sea will have to be tackled internationally— better it had been prevented altogether.

Finally we must decide how we want to live, and the kind of decisions we make will influence the future and the lives of our descendants. If we have any hopes left we can realize them only if we know what we are doing and what we have. If the situation is left to drift, we shall in a very short time exhaust the exploitable resources on the one hand and on the other completely destroy our environment. Having decided to launch our great rescue operation, using all the technology that we can, having put the vast oceans and the atmosphere under international control, having cleaned up our own immediate habitat, we must then decide what kind of habitat we want to live in, bearing in mind all the time that our numbers will dictate the possibilities available.

If we decide to protect and leave vast tracts of countryside undeveloped (and this means rejecting the idea of the mega-polis), then for some years we shall have to build up or down. We have already experienced the failure in social terms of high-rise flats which have been responsible in many cases for destroying relationships between people, have deprived children until school age of social intercourse and condemned the aged members of our community in their last years to utter loneliness. The techno-architects are concerned only with space and volume and not with intelligence. Techno-architecture, or 'arcology', the so-called combination of architecture with ecology, presents us with a total planned environment: houses, factories, schools, universities, etc., built within a single megastructure, 300 storeys high and two to three miles wide. This is not a dream; such megastructures have already been designed in the United States, the ultimate in habitat with inhabitants completely divorced from nature, air fed to their quarters through snorkels, wind speeds, temperature and light intensities controlled by computers. This may enable us to survive after a fashion, and may tempt us to leave the decision about our numbers to other generations, but if our species is to survive we have to make the decision in *this* generation and work hard to achieve our goals. This is an ultimatum for Mankind's survival.

CHAPTER 17

The outlook for *Homo sapiens*

In the summer of 1971 it is questionable whether mankind has a future. We have examined only part of the problem that faces our species in this book, but we have tried to put it into the context of the overriding, seemingly insoluble problem: there are too many of us.

If we could only solve this problem we should be well on the way to solving the others. Everything hinges on the fact that we are making excessive demands on our environment—further, we may say that we are making too many demands on ourselves. For while we are benefiting on the one hand by our material advances, we are losing much that goes to make up the quality of life on the other. It seems that we are on a countdown to disaster, whereas we could, by using sense and restraint, be riding a road which befits an organism that has called itself *Homo sapiens*. The exponential curve of technological change is now approaching the vertical. There are few people who, having given thought to this, do not believe that at some time the crash will come. The only area of disagreement is when it will occur. The longer it takes, the worse it will be; the more people there are alive when it does occur, the greater will be the disaster. But we, in this second half of the twentieth century, must not get into the nasty little habit of thinking that it will not happen in our time. Breakdowns in our society are already occurring and the quality of life leaves much to be desired. The situation is rapidly approaching the flash point of an environmental implosion.

We may still have time, we may still have a choice, but it seems to me only just. If we go on as we are, maintaining the present rate of growth in the human population and demanding so much of the environment, we shall inevitably run out of resources and life will cease to exist on the planet. It is possible that Man will adapt to and survive for a time in an increasingly deteriorating environment; in time the new regime induced by the pollution and waste factors will take the place of the old environmental filters that once ensured that only the fittest survived. In other words the men and women of the future will be the 'fittest' to meet the demands of the industrialized environment. The conditions in which these 'creatures' will live will be a long way from the ideals to which the human mind of today generally aspires. There will be no countryside, no wild life, no clean and refreshing oceans or cool breezes on a summer's day. Instead there will be a shining steel and concrete googleopolis, the sea a stagnant and evil-smelling fluid, the total environment a mono-culture. Man will have become like bacteria, but living in a petri-dish of his *own* making. Escape will then be found only in the hallucinogenic drugs and gadgetry of the scientists. And Man's end will come when he can no longer support the machine. The fabric of his artificial world will crumble when the machine stops.

If we do not adapt, and the present trends and changes in the environment continue until even with our technology we cannot call a halt to its downward progression, then the exhaustion of the Earth's resources, the depletion of the oxygen supply, the degenerative processes will bring the great biological experiment to an end.

We have one more choice, one to which most thinking people —at least, so I hope—would subscribe, and that is to create a life based on quality and not quantity. Primitive Man was totally involved in the struggle to survive; when, later, his creative genius enabled him to achieve technological mastery it marched hand in hand with his blind destructive nature. Now he has arrived at the point at which he has to take stock, and he finds himself wanting. Before, he looked for outside influences to help him in times of crisis or doubt—he invented religions, philosophies and political ideals; in the years since Hiroshima and Nagasaki he has been desperately searching for some unifying influence to replace these forgotten 'hopes', but his political ideologies and selfishness have thwarted his search. But he can no longer afford to postpone the day of decision; he has to choose the kind of life he wants for himself and for those who follow him. There is a goal, one that has the potential to unite every man, woman and child on this planet, which, if reached, will enable them to build at last that 'land fit for heroes'. A home which they can be proud they helped to build, one in which they could live in harmony with the wild things, retaining the beauty of the mountains, the lakes, the rivers, the fields and the vast oceans and the sky. But this needs effort and requires a genuine desire on the part of every one of us to make this dream reality. Politicians who have the effrontery to believe they know what is good for us must be made to work out the paths which we must follow to obtain the quality of life which has its source in our highest aspirations.

It seems to me at this time highly improbable that this last choice will be taken. There are numerous reasons that lead to this conclusion. Ideals such as this have been formulated before and have never been reached. The psychological, behavioural and anthropological sciences give little hope that human nature will change, and this choice requires above all else a change of heart, if not of mind. If it is true that we get the environment and conditions we deserve then perhaps we have already created the tomb in which we are to die.

The sad thing about all this is that the conditions we are creating now will affect those who follow us, and although they may have higher ideals than we have and be willing to make sacrifices and to work harder to build a better habitat, because of what we are doing now they will never be able to attain their dreams. As I have said already, the longer the problem remains unsolved, the more complex it will become and the less able will our technology be to pull us out. It is criminal to say that the future will take care of itself; it cannot. Each one of us, the writer of this book and his reader, plays his part in this crucial time. It is at our level, the grass roots as it were, that the problems begin. It is up to us to regulate our numbers. This needs discipline. It is in our interests and those of our descendants that we have only enough children to replace ourselves and to provide them with an environment which will enable them to grow up in a free and healthy society.

We cannot afford our excesses. Even if we had unlimited resources and room we should still have to be wary of where innovations were leading us. As it is, technologists are leading us nowhere fast because we are already at the limits of what society, and in particular the individual human being, can sensibly utilize at any one time. We have sufficient knowledge and techniques available to sort out the problems that confront us now. We must begin to look in those areas which have real problems, those which are affecting our lives, our future and our survival. We must persuade the talented members of the community to tackle the problems of the community. If it is true that the proper study of mankind is Man, then the proper interest of mankind is his environment; we are part of it and we are creating a new habitat out of the old. But what we have managed so far is not worthy of us. We have not much time to stand and look, we have got to think hard and act.

Paradoxically it is our technology that provides the answer. Not so long ago three members of our species were all but lost in space: 50,000 scientists and technologists combined their efforts with the prayers of millions of ordinary people on this planet to bring them back alive. We are all on a space ship hurtling through the black emptiness of space. We cannot go back anywhere, we have got to make the best of what we have, and it is going to take all the talent, all the genius and all the prayers of mankind to keep it going.

POSTSCRIPT

The reader of this book may ask how he as an individual can influence events which will undoubtedly affect him and certainly his descendants. At times in this age in which we live, it seems that questions of this nature are futile, but we in the West are not quite as helpless as we sometimes think. We have the vote and we can use this to good cause if we wish it. The thing we have to do first of all is to get out of our rigid attitudes to a political system which demands allegiance to one party of political thought or another. Instead we should ascertain what the individual politician who is standing for election in our area is going to do to ensure that our environment is safeguarded. If he has a positive approach and seems likely to make this issue his foremost task then we should give him our vote regardless of his political colour. If he fails us we should root him out at the first opportunity.

We should not tolerate a political machine that gives lip service to the solving of the fundamental problems described in this book while at the same time doing little to put the situation right.

This use of the vote is our strongest weapon and is just as valid in local elections as in national elections. If we want to take a more active part in the day-to-day running of our environment we can join one or other of the local societies or national organizations devoted to environmental issues. If we are young and able-bodied we can join the Conservation Corps and help in our spare time in the numerous tasks which can only be done with this voluntary help. In the last resort it is we as individuals who will produce the environment we deserve. If we think we should have better then we have to work for it.

BIBLIOGRAPHY

Further Reading

For those readers who wish to pursue the arguments developed in this book in more detail, the following books may prove helpful. Although their authors may not always agree with the views of the writer of this book, it does not detract from their value, because the problems of which we are writing are so complex that more than one viewpoint is valid and necessary.

GENERAL

Survival: Man and His Environment
Don R. Arthur; English Universities Press, London, 1969
Before Nature Dies
Jean Dorst; Collins, London, 1970
Nature Conservation In Britain
Sir Dudley Stamp; Collins, London, 1969
Man and Environment: Crisis and the Strategy of Choice
Robert Arvill; Penguin Books, Harmondsworth, 1967
Blueprint for Survival
Robert Allen and Edward Goldsmith: *The Ecologist*, Vol. 2 No. 1, January 1972.

SPECIFIC

The City in History
Lewis Mumford; Penguin Books, Harmondsworth, 1966
Pesticides and Pollution in Britain
Kenneth Mellanby; Collins, London, 1967
The Waste Makers
Vance Packard; Penguin Books, Harmondsworth, 1963
Derelict Britain
John Barr; Penguin Books, Harmondsworth, 1969
The Last Resource: Man's Exploitation of the Sea
Tony Loftas; Hamish Hamilton, London, 1969

Drugs: Medical, Psychological and Social Facts
Peter Laurie; Penguin Books, Harmondsworth, 1969
Drugs and the Mind
Robert de Ropp; Gollancz, London, 1958
The Ultimate Folly
Richard D. McCarthy; Victor Gollancz Ltd, London, 1970
Chemical and Biological Warfare
Seymour M. Hersh; Panther Books, London, 1970

In addition to the above there are the reports in the press and in particular the weeklies. There is no one journal in the United Kingdom which covers the subject of pollution in its entirety, but *New Scientist* and *Nature* are notable for regular reports of different aspects of the problem. In the United States they are more fortunate in that they have an excellent journal, *Environment*, which attempts to cover the problem— unfortunately this is not yet available in the United Kingdom except by direct subscription to the publishers. For government views readers should consult the occasional official reports such as the *Royal Commission on Environmental Pollution: first Report*, H.M.S.O., 1971; and of particular note is *The First Annual Report of the Council on Environmental Quality*, published by the U.S. Government Printing Office in 1970; *The Second Annual Report of the Council on Environmental Quality* was published 1971.

INDEX